PENGUIN BOOKS

Concorde

Mike Bannister is an aviation consultant and former pilot. He became the youngest pilot on the Concorde fleet in 1977, was appointed British Airways' Chief Concorde Pilot in 1995 and regularly flew as captain on all of the aircraft's routes worldwide. He was at the controls when BA's supersonic flagship returned to service in November 2001, and commanded and flew the final Concorde commercial flight from New York to London on 24th October 2003. By the time Concorde retired, Mike had amassed over 9,200 Concorde flight hours – around 6,900 at supersonic speeds. For over ten years he was extensively involved as an expert witness and lead technical advisor in the 'Air France Concorde Trial'. He now runs an aviation consultancy and resides in Middlesex, UK.

Concorde

MIKE BANNISTER

PENGUIN BOOKS

PENGUIN BOOKS

UK | USA | Canada | Ireland | Australia
India | New Zealand | South Africa

Penguin Books is part of the Penguin Random House group of companies
whose addresses can be found at global.penguinrandomhouse.com

First published by Penguin Michael Joseph 2022
Published in Penguin Books 2023

001

Set in 12.02/14.24pt Garamond MT Std
Typeset by Jouve (UK), Milton Keynes
Printed and bound in Great Britain by Clays Ltd, Elcograf S.p.A.

The authorized representative in the EEA is Penguin Random House Ireland,
Morrison Chambers, 32 Nassau Street, Dublin D02 YH68

A CIP catalogue record for this book is available from the British Library

ISBN: 978-1-405-95192-0

www.greenpenguin.co.uk

To my very long-suffering family,
who I took to 'The Edge of Space'

Contents

Prologue

25th July 2000

The starboard flank of the *QE2* soared above us. Some confusion over our sailing time meant we'd arrived at Southampton Docks an hour late. Our bags were in our cabin; the ship was about to sail.

We were halfway up the gangplank when my pager went off and my mobile phone rang at the same moment. I glanced at the message on the pager: 'Call BA. Most urgent.'

I was patched through in seconds to the duty manager at our Heathrow Emergency Centre.

'Mike, I've got terrible news,' he said. 'There's been a crash – a Concorde crash.'

The details were sparse and coming in as we spoke. Air France Flight 4590 had slammed into a hotel on the edge of a small town called Gonesse, 2 miles from the end of runway 26R at Paris's Charles de Gaulle airport.

My wife Chris, a former cabin service director, was part of the BA family too. She immediately knew how serious it was – and didn't need me to tell her what this meant for our much-anticipated Atlantic crossing. Our daughter Amy had been so excited she hadn't talked about anything else for months.

As soon as I hung up, she touched my arm. 'Don't worry about Amy or me, Mike. Just do what's right.'

I called the driver. Within a beat, he'd turned around and was heading back to the Ocean Terminal.

Chris, meanwhile, spoke to the Cunard check-in desk about rescuing our bags, whilst trying to explain things to one very disappointed six-year-old.

While waiting for the car to arrive, I called my counterpart in Air France. In the background, as he picked up, I could hear sirens – the distinctive, discordant wail of French emergency vehicles.

Curiously, as the only two airlines to operate the world's most exclusive passenger plane, there hadn't been more than cursory contact between BA and Air France Concorde crews for many years. When I became BA's Chief Concorde Pilot in 1995, I made an effort to put that right. Edgar Chillaud and I were still getting to know each other, but I already respected and liked him a lot. The fact that my opposite number at Air France was now at the crash site shocked me. I'd assumed he'd be at their equivalent of our crisis centre.

'Everything, it is burning,' he said. 'It is terrible, Mike. Truly terrible . . .'

I mumbled something about the possibility of survivors.

'I don't think there could be.' Most of the wreckage, and thus the people on board, were contained within the impact site.

He told me the name of the pilot: Christian Marty, a Concorde veteran. Captain Marty had made a name for himself in 1981 as the first person to windsurf across the Atlantic.

Our Concorde crews were a close-knit family; Air France's, being smaller, were even closer. My presence on the call suddenly felt like an intrusion. I could only imagine Edgar's pain and grief as he confronted the full horror of the disaster.

As the car pulled up beside us, Chris and I agreed that she

and Amy would go home and the driver would then take me to Heathrow. I climbed in beside him and had hardly fastened my seatbelt before he said: 'Have you heard? There's been a terrible Concorde crash. Everyone killed.'

Without a pause, he proceeded to tell me what had caused it and that it was 'the end of that beautiful aircraft'. I said nothing because I was already patched into a conference call with David Hyde, BA's director of safety and security, Alan McDonald, the airline's director of engineering, and Geoff Want, general manager of operations. For a while, we were joined by Rod Eddington, the chief executive.

We went through the options. At one end of the spectrum, we could continue as normal – this was an Air France accident; we should maintain a stiff upper lip and carry on. At the other was the nuclear alternative: halting all Concorde operations indefinitely. Continuing as normal got my vote. As tragic as this was, I said, if we responded purely emotionally, it would be difficult, if not impossible, to come back from it. Emotions, unlike facts, were hard to push against; facts we could deal with.

At the back of my mind was the knowledge that Concorde had become a Marmite issue inside BA. For many of the senior execs, she was 'love it or hate it'. A number of them, including some now on the call, saw it as a massive drain on the airline's resources and had been looking for excuses to kill it. Others – me amongst them – saw it as a huge asset. Until this moment, the aircraft had operated for twenty-four years largely without incident or interruption and there was no reason why she shouldn't carry on – unless the evidence told us otherwise.

I was still on the call when we dropped off Chris and Amy. The Emergency Centre was a few miles up the road. Its interior reminded me of the bridge of the Starship *Enterprise*, with TV screens on the walls and banks of computer monitors on its many desks. Profound shock and sorrow were etched on every face as I walked in.

It was now coming up to 18:00.

A series of nightmarish images had started to appear on one of the giant screens. A still shot by someone close to the end of Runway 26R showed a massive jet of flame streaking from the Concorde's left wing, just inboard of the two engines, moments after take-off.

Video footage taken on a camcorder by a passenger on a passing truck tracked the stricken aircraft for several more seconds. It was almost unbearable to watch: her back end now ablaze, a trail of smoke darkening the sky.

Air France had immediately cancelled all its Concorde operations for the duration. As a mark of respect, we decided that we would cancel our next one, Flight BA003 to JFK, scheduled to depart in a little over an hour.

As the night wore on, we coordinated our strategy in response to the emerging details.

The aircraft had been chartered by a German company, Peter Deilmann Cruises.

The one hundred passengers had been on their way to board the cruise ship MS *Deutschland* in New York City. From there, they would have steamed south on a 16-day cruise to Ecuador.

The flight had been running late.

Metallic debris on the runway had blown a tyre.

Fragments of the tyre had cannoned into the underside of the port wing, causing an A4-sized hole in the fuel tank between the Number 2 Engine and the fuselage.

News from Paris indicated that the French authorities

were already zeroing in on a titanium alloy wear-strip that had apparently dropped off the engine reverse-thrust mechanism of a Continental Airlines DC-10-30, which had left for Newark five minutes earlier.

But on its own, I told myself, this should never have caused the disaster.

What in God's name had gone wrong?

It was midnight by the time I got home, shattered from the day's events.

Chris told me that Amy had seen the news, that she understood what had happened and that her disappointment over our cancelled New York trip had been replaced by her terrible sadness for all the people who had died.

The next morning, I arrived at the Emergency Centre at 06:00 to learn that almost all talk of the accident now revolved around the wear-strip. When a complex piece of engineering fails, it is predominantly because a causal chain of events – a whole host of factors – has led to a system-wide failure. All our questions now hinged on the extent to which Concorde's failure had been mechanical, human, or both.

I joined the crew of BA001 before they left for New York to tell them everything we knew – and, most importantly, that in our assessment, Concorde was safe. They needed little persuading. Only one of the cabin staff chose not to fly. The passengers had remained loyal, too. Barely a handful had cancelled.

After another gruelling day at the Emergency Centre, I got home just as Amy was going to bed.

I asked if she knew where I had been.

'Yes, Daddy,' she said quietly. 'Making sure it never happens again.'

That night, as my head hit the pillow, I had little inkling of the forces that were building even then to dismiss the complex causes theory of the crash.

Or the degree to which some people wanted to bury the aircraft along with the victims at Gonesse.

Or the fight I faced to save her reputation.

As I lapsed into a fitful sleep, a part of my mind flew westward, streaking at Mach 2 over the Atlantic.

Faster than a rifle bullet, 23 miles a minute.

Faster than the Earth rotates.

Buying back time.

I was once again at the controls of an aeroplane that could outpace the Sun – from an altitude where you could see the curvature of the Earth.

Somewhere in my dream I looked up. I was my seven-year-old self again, gazing at the stars and within touching distance of space.

I

Bournemouth Beach, a beautiful day in that otherwise wet summer of 1956.

I am at the water's edge, admiring the sandcastle I've just made. I turn to look across a millpond-calm sea at a pier stretching into the middle-distance. A handful of swimmers are testing the temperature of the water. I look up at the blue, blue sky and enjoy the warmth of the sun on my face. I glance back at Mum and Dad. Mum waves, while Dad just stares out to sea.

The year before, when I was six, he lost his job. My father had been the last trainee cooper in England.

For a year or so, there had been talk we might emigrate to Tanganyika from my home town of Burton-on-Trent, but we ended up moving instead to Luton, where my dad had been offered a job as a dairyman. We didn't own a car, so my horizons were limited, expanding briefly when we went to stay with friends in Southbourne, near Bournemouth, for our traditional summer holiday.

The coach journey took five-and-a-half hours and I loathed it.

Which is why, on that beach, I spent my summers looking up.

In those days, the mid-to-late fifties, I could identify every speck of silver against the azure.

They were mostly piston twins or turboprops: Douglas DC-3s, triple-fin Airspeed Ambassadors and sleek Vickers Viscounts.

Where were they going? France? Italy? Somewhere more exotic?

The most important thing – the thing that mattered most on those beach holidays – was that a plane would get me where I wanted to go a lot faster than that old coach.

On that rare hot day that summer, after a week of staring skyward, I tottered up the beach with my bucket of shrimps and announced solemnly to my mum and dad that I was going to be a pilot.

'Yes, dear,' my mum said, before turning back to her book.

The years passed, but I never lost sight of my dream. I knew what I wanted to be. After O levels, the time came to sit down with my careers master and tell him. He stared at me: no pupil of the school had ever chosen *flying* for a career. What about the civil service, industry . . . teaching, even?

That same day, I went to the library and found out what A-level subjects I needed.

As I approached the exams, I thought it might be a good idea to have a back-up plan and managed to get a conditional place at university to read aeronautical engineering. However, my dream was still flying. I didn't much care whether I joined the RAF or an airline at that point – all I wanted was to get behind the controls of a plane.

My research had shown me that learning to fly for a career in the RAF or the airlines was subsidised. If I joined the RAF, they would pay *me* to learn; if I went to Hamble College of Air Training, where pilots for the British Overseas Airways Corporation and British European Airways received their training, I would be eligible for a £2,000 grant. We had to pay for our keep, but the Hamble course itself was paid for by BOAC, BEA and the government.

In the summer of 1966, hedging my bets, I applied to the RAF and to Hamble. The RAF assessment at Biggin Hill in Kent was the first to come up.

I was pretty good at mental arithmetic and non-verbal reasoning – two of the skills they would both be looking for – so felt reasonably confident about the exam side of things. Beyond this, I knew the Air Force needed officers with leadership skills, while Hamble was focused more on aptitude, determination and a desire to follow a long-term career in aviation.

At Biggin, I finished my exams early and, feeling rather pleased with myself, handed my paper in. The invigilator asked if I was feeling all right.

'Perfectly, yes, thank you.'

He tapped his watch and tutted. 'You're not supposed to finish early. The mental arithmetic and non-verbal reasoning papers are set so you shouldn't have enough time to finish them.'

Still pondering his reply, when my name got called out a short while later, I was already convinced that it was to tell me I was being sent home.

I was led into the office of a group captain, who looked like he'd clambered straight out of a Spitfire. He glanced up as I stood in front of his desk.

'Bannister?'

'Yes, Sir.'

'Take a seat. We've been reviewing your assessments.'

My heart sank.

And then he said: 'Have you thought of becoming a navigator?'

'A navigator?' I stumbled, cursing myself for handing those papers in.

9

'I really want to be a pilot, Sir, but if that's all you're able to offer me, of course I'll consider it.'

'No, no,' he said, smoothing one side of his moustache as he glanced up. 'It's just that you excelled in mental arithmetic and non-verbal reasoning and we're looking for navigators and think you'd be an exceptional candidate.'

I said: 'That's kind of you, Sir, but I really, really want to be a pilot.'

He stared at me for a moment longer, muttered something that didn't exactly sound like approbation and pulled open a drawer. He riffled through some papers, found what he was looking for and pushed a white form at me. 'Very well,' he said, 'have you ever thought of applying for one of these?'

I stared at the jumble of words. 'One of what, Sir?'

'A special flying award.'

'I'm sorry, but I don't know what that is.'

'Well, basically, if you apply for it and you're successful, the RAF will sponsor you for pilot training up to PPL standard at a civil training school.'

I didn't need to be told twice. 'Yes, thank you. I'd very much like to apply.'

He nodded. 'Good. Fill it in, then.' He pushed the form closer.

It took me around fifteen minutes. When I pushed it back across the desk, he looked up and beamed: 'I'm pleased to be able to tell you, Bannister, that you've been given a special flying award.' He shook my hand. 'Good show.'

A few weeks later, I sat my exams at Hamble and was offered a place on a course there as well.

But I had also been offered a place at the RAF Air Training

College at Cranwell – and by now my heart was set on a career in the Royal Air Force.

In the late sixties, Luton Airport had become home to a number of airlines that were beginning to grow thanks to the recent boom in the package holiday sector. The biggest of them was Britannia Airways, which had recently been equipped with the Bristol Britannia, a magnificent British-designed, four-engine turboprop, by now, mercifully, past its early teething problems.

It was 1966, and I was just past my seventeenth birthday. Buoyed up with funding from my award, I began to train for my PPL at the airport.

My dad had become the manager of a large Co-op warehouse in Luton (my mum also worked for the Co-op). We lived on-site in a comfortable flat that came with the job. The airport was almost on our doorstep.

For four weeks of that summer – a year before I took my A levels – I wound my way up the hill to the main gate, the air becoming thicker with the heady tang of burnt jet fuel the closer I got.

My instructor, Bob, was a big man who also worked at Luton as an air traffic controller. He had coached me to within a flight or two of my first solo in a Cessna 150. The number of additional flights that would get me there would now be determined by how I fared on this one – a simulated engine-out at 3,000 feet.

Managing an engine failure soon after take-off is a critical issue for any pilot. Simulating a failure on a single-engine, twin-seat Cessna 150 involves closing the throttle, assessing why the engine has stopped and trying to get it re-started, all the while scanning the terrain below for somewhere to land.

You're looking for something that's flat, without obstacles – cables and electricity pylons especially – but the fields that look the most appealing are quite often under cultivation. From 1,500 feet, though, you can normally tell the difference between a cricket pitch and a field of cabbages. And by 400 feet, on your glide in, the instructor can usually tell if you'd have made it or not.

On this particular day, as we took off, I'd already blown the exercise once – I'd have landed, Bob acknowledged, but it would have been rough.

'Let's have another go,' he said patiently, as we climbed back up to 3,000 feet. 'I'm detecting some over-confidence, Michael.'

He'd barely finished the sentence when he cut the engine.

The point of a simulated engine failure is that it's supposed to be simulated. At no point was Bob meant to switch the engine *off*.

With the din of our flat-four Continental suddenly replaced by the whistle of our slipstream, I appreciated in a flash just how big Bob was.

Our little Cessna had started to fall like a rock.

I quickly trimmed the controls and established glide speed, all the while searching for a suitable field to land. I spotted one on the edge of town and started to make for it. Meantime, I went through the emergency re-start drill.

At 150 feet, far lower than you'd normally call off a simulated engine-out, Bob announced himself satisfied and restarted the engine – and never had that noise or the power of a hundred horses sounded so good.

After lunch, we did a few more circuits and bumps before touching down for what I thought was the end of the day. But Bob said, 'See you in twenty minutes,' and got out.

Strolling away from the aircraft, he shouted over his shoulder: 'Your first solo, Michael. Enjoy it.'

Shed of Bob's 14 stone, the Cessna seemed to take to the sky like a rocket.

As I flew among the clouds, knowing Bob, back in air traffic, was watching and listening, I felt, for a few short moments, like I was all my aviation heroes rolled into one.

The next summer, I was all packed up, travel warrants in hand, a week from heading off to RAF Cranwell, when my father fell ill. The local doctors were utterly stumped – all they knew was it was a viral infection of some kind.

For a few days, we watched, helplessly, as, swollen and bed-ridden, my dad lapsed into a sickness from which we didn't think he'd recover.

I called the RAF and informed them what had happened, and they were brilliant – not a problem, they told me, attend to your family and we'll see you next year instead. They reserved a place for me on the March '68 course.

An ambulance arrived and took Dad to a three-bed intensive care unit at St George's Hospital in South London. Mum, my brother Keith and I followed. Dad remained in specialist care for three weeks and survived – the only person in the unit to do so.

As my thoughts drifted back to the RAF, I got a phone call out of the blue from Hamble. It was the third week of November 1967. Somebody had dropped off the course, the voice on the other end of the line told me – they had a spare place *right now*. Did I want it? Within 30 minutes, I'd said yes.

2

A few days later, I arrived at a grass airfield, waterlogged from a week's worth of rain.

With a sharp, wintry wind blowing in off the Solent, Hamble felt like it was an unlikely place for the cream of Britain's next generation of civilian pilots to be commencing their careers – preparing to handle a new breed of jets like the Boeing 707, Vickers VC10 and Hawker Siddeley Trident.

But '67 was an unusual period in the history of UK civil aviation.

Hamble had been set up in 1960 as an exclusive training establishment for BOAC and BEA. BOAC flew intercontinental routes; BEA served the UK, destinations across Europe and parts of the Middle East. Eventually, they would come together to form British Airways. In 1967, both were facing a pilot shortage – something that would have been unthinkable a few years earlier.

For fifteen years, Britain's civil aircrew needs had been furnished by ex-military pilots who'd drifted into the airlines after the war. Cutbacks in Britain's armed forces combined with a wave of retirement amongst the wartime generation of pilots flying for BOAC and BEA had quickly turned a glut of aircrew into a drought. To bring a new generation of civil pilots into the air, BOAC, BEA and the Ministry of Aviation pooled together to form Hamble.

Four streams of forty-five students per course produced a theoretical output of 180 airline-ready pilots per year. But the two-year syllabus had recently been condensed into eighteen months, with just two weeks off each academic year – a week at Christmas, another at Easter. The rest of the time it was work, and the drop-out rate was vicious.

By the end of those eighteen months, a third of us were expected to have failed.

My group of fifteen (a third of the stream that started that November) – 'C Course' – was tailored towards those who weren't as strong at maths and science. Had I been able to join in the normal way, I would have slotted into a group that was good at those subjects, because I was naturally good at them, but it didn't work out that way, and I wasn't complaining – I was just thrilled to be here.

Hamble was a mix of new buildings and old – the old mainly being prefabs put up during the war. Our course was grouped in one of the older blocks, which wasn't all bad, because the rooms, though draughty, were bigger than those in the newer buildings.

My fellow students spanned the age range of the entry requirement: eighteen to twenty-six. I was the youngest on the course and found myself sharing a room with two of the oldest. Tom had come to Hamble from the Navy, where he'd been a radar officer on the aircraft carrier HMS *Victorious*. Roger had no experience of aeroplanes at all.

Tom had a girlfriend and money in the bank. He also owned a sporty little Mini Cooper S. Together with another member of the course, Anthony, who'd been an insurance salesman and owned a flat in London, we became a tight-knit group – my place cemented, initially at least, by the fact I could help them with maths and science – Mr Popular, by default, in a group that struggled with those subjects.

The first six weeks were spent in the classroom learning

the basics: how piston- and gas-turbine engines worked, aerodynamics, basic navigation.

The next part was a combination of technical and flying. *Ab initio* was carried out on the de Havilland Canada Chipmunk, a pretty little post-Second World War tandem-seater. Next, we went on to the Piper Cherokee for instrument flying and early-stage cross-country navigation – day trips to airfields around the country with rapid turnarounds back to Hamble – learning the rudiments of what it would be like in an airline to fly on a schedule. Only if we passed a final written exam three-quarters of the way through the course would we get to transition to the 'advanced' part of the syllabus – flying the twin-engine Beech Baron in airways and in and out of major airports. And provided we got through *that* all right, we would emerge with a CPL – a commercial pilot licence – as well as an instrument rating and a guaranteed job at either BOAC or BEA.

I was swotting for my finals in the study room one spring evening with the black-and-white TV on in the background when a newsflash reported that, five weeks after the first flight of the French-built prototype, the maiden flight of the first UK-built Concorde, Concorde 002, was taking place from Filton, near Bristol, and the BBC was covering it live. Forget the books, I was now gripped by the telly.

Concorde 002 landed twenty-two minutes after it had taken off at RAF Fairford in Gloucestershire, from where it was due to undertake its operational trials. The veteran BBC journalist Raymond Baxter, a former Spitfire pilot, was there to cover the event. Evident in his voice was his own pride in this achievement.

Test pilot Brian Trubshaw, he told us excitedly, had decided to lift 002 into the air on what was supposed to be just another high-speed taxi-run.

On landing, 'Trubbie' told reporters that the flight had been 'cool, calm and collected' and that the six-man crew had enjoyed the flight immensely.

Up until that moment, the only aircraft in my sights had been the VC10 – what I considered to be the most beautiful plane ever built. With its high T-tail, the VC10 was both graceful and distinctive – and it was British.

But then there was Concorde. Who, in Britain, hadn't followed the progress of this extraordinary aircraft?

Developed jointly with France, launched in 1962, rolled out in 1967, and first flown by the French a few weeks earlier, the plane had already cemented itself into aviation lore as an icon. With years of testing ahead of it, even after this milestone event, I had never until this moment considered it to be an aircraft that *I* might fly one day. But seeing it now, listening to Raymond Baxter, and revelling in Trubbie's cool, test-pilot *sang froid* as much as the veteran reporter was, I felt a sudden and strange feeling of anticipation.

One day, I told myself, I was going to fly that aircraft, too.

Statistics don't lie – by our final exam, one-third of the students out of the forty-five who'd started our course had been culled.

Being less science- and maths-oriented, our course had taken a bigger hit than the others.

Then we were hit by a bombshell.

Our principal told us that, due to a slowdown in economic demand, which had rippled through to the airlines, neither BOAC nor BEA would be taking us on as pilots.

To say this was a blow was an understatement. When I'd stared up at those shimmering silver specks from Bournemouth Beach, the dream hadn't been about a career as a *navigator*. But that was the only one now on offer.

Over the next few weeks on the last part of the course – the period when we were training on the twin-engine Baron – we were confronted by an unusually long period of bad weather and the pace of advanced training slowed.

Our graduation day was set in stone – with groups rotating through on a fixed schedule, there was no room to extend it by more than a few days.

To graduate, you had to pass the course *and* have your CPL and instrument rating. If you hadn't notched them up by then, you weren't allowed to attend graduation.

It was touch and go – but on the morning of the ceremony, I was able to get into the air and was informed that I'd passed. This appeared to come as a considerable relief to my instructor for reasons that only became clear after we landed, when he marched me off to see the principal.

'Ah, Mike,' he said, as I closed the door behind me. 'I'm happy to tell you that you've graduated top of your course, and of this whole year's 180 student intake.' He shook my hand warmly.

Pleased as I was, my reaction was tempered by the knowledge neither BOAC nor BEA wanted us.

'I'm especially glad you're graduating today,' the principal continued, 'because I'd like you to make a speech at tonight's dinner.'

So, that's why my instructor had been relieved.

I thanked him and said I'd see him at the dinner.

After my speech that night, the principal got to his feet. 'The bad news, as you know,' he said, 'is that neither BEA nor BOAC want you as pilots – and there's been no change there. The good news is they're still committed to giving you the job you were promised when you started this course. And something else. They've said they will honour your preference of airline. As you know, this has never happened before. I hope, therefore, this will come as some consolation.'

Every previous student at Hamble had merely been able to state their preference as to which airline they wanted to join, with no guarantee they'd get the one that they'd asked for. I reasoned that recessions didn't last forever, and because long-haul felt more challenging and glamorous than short-haul, I opted for BOAC; as did two-thirds of my fellow graduates.

One of the things I'd enjoyed most about the camaraderie at Hamble was the degree to which we'd helped coach each other through.

But for Anthony, the writing had been on the wall for a while.

We all went down to the pub and got a little pissed before saying our goodbyes.

Within a few weeks, Anthony would be back in his old insurance job. For the rest of us, even though we knew we were lucky to have jobs, the future felt a lot less certain.

3

A couple of weeks after Hamble, beneath a flat, slate-grey sky, I arrived at BOAC's Heathrow Training Centre at Cranebank, a five-storey block with a flat roof and a drab façade. I glanced up as I approached the entrance and saw a blue disc with a plaque beneath it. The girl on reception, who'd seen me looking at it, smiled and said: 'That's the height of the fin of one of our new Jumbos.'

The 747 'Jumbo Jet', as it had been nicknamed, first of the new 'wide-bodies', would join BOAC the following year.

Even the arrival of the mighty Jumbo wasn't going to help me.

Since we'd left Hamble, BOAC had informed all its would-be cadet pilots that, with no pilot places available, we were going to be trained as 707 navigators.

In our era of satnav, it's easy to forget that until relatively recently, four-engine airliners had real navigators. In the sixties, it was even more basic. Even twenty-four years after the Second World War, there were no trans-oceanic navigational aids. 707 and VC10 navigators – emulating what had been done by navigators for centuries – stuck a sextant through a pressure air lock in the flight deck roof to shoot the stars, the Sun, the Moon and Venus. This ensured that the aircraft – blown by unpredictable winds and guided by instruments that weren't as accurate as today's – stayed on course.

For two weeks, we sat in a classroom at Cranebank listening to an induction officer telling us about the airline we were now serving. The previous October, BOAC had introduced non-stop services to Antigua and Barbados. It had also recently opened a new computer complex at Boadicea House within the Heathrow complex. A week or so previously, a new 'Polar Route' had been inaugurated: London to Anchorage and on to Tokyo and Osaka.

Next up, 'Safety Equipment and Procedures': how to evacuate an aircraft, where to locate the emergency equipment, how to carry out CPR, first aid, and so on. Our attentiveness had been assured by a shocking accident that had occurred on a BOAC 707 flying out of Heathrow the year before, the details all too fresh in my mind.

Speedbird 712 – *Whiskey Echo* – had taken off from Runway 28L with 127 people on board, destination Sydney via Zurich and Singapore. The crew included the addition of an acting first officer, John Hutchinson, later to become a Concorde pilot and friend, and a check captain, Geoffrey Moss. Seconds after take-off, there was an enormous bang and the aircraft started vibrating. Moss looked out the window and yelled that the wing was on fire. Two minutes after take-off, the crew called a Mayday.

As the aircraft turned back for an emergency landing on Heathrow's Runway 05R, the Number 2 Engine *fell off* – the fire having melted through the pylon. It hit the ground in the nearby village of Thorpe. Fortunately, nobody there was hurt.

Not so, tragically, those on Speedbird 712. Upon landing, the cabin crew initiated an emergency evacuation, but the conflagration by now was so bad that the port wing exploded before the fuel shut-off handles could be activated. Thirty-five people were injured; five were killed – including a stewardess who'd gone back into the aircraft in a bid to rescue the four passengers still inside. For this, she was post-humously awarded the George Cross.

In the ensuing investigation, it was found that metal fatigue in one of the compressors of the Number 2 Engine had set off a chain of failures.

These had included the severing of the main fuel pipe, damage to the engine's two extinguishers and the crew's omitting to shut off fuel to the engine – kerosene from the booster pump intensifying the fire until it spread to the wing itself. On landing, the fire spread under the aircraft and ignited the fuel lines and oxygen tanks, causing a series of explosions that burst into the cabin.

Two weeks into the course, a senior management pilot walked into our classroom, wearing his captain's uniform, which was unusual – perhaps, I thought, he'd come straight off one of the simulators, which were in another part of the building.

After a whispered conversation with our classroom instructor, our distinguished-looking visitor turned to us and said: 'Right, gentlemen, I'm going to pass a form round. I'd like you to fill it in. You'll see there are some boxes. All you have to do is tick one of them, then hand the form back to me.'

On the form were five questions: 1. *I very much want to be a 707 pilot.* 2. *I want to be a 707 pilot.* 3. *I don't mind if I become a 707 or a VC10 pilot.* 4. *I very much want to be a VC10 pilot.* 5. *I don't mind being a VC10 pilot.*

I went with my instincts and ticked Box 4, because nothing had changed: I still wanted to fly the VC10.

Six of us ticked that box. The rest, we learnt later, sensing it was a trick, ticked Box 1. We were, so their logic went, learning to be navigators on the 707, after all; so that, they reasoned, was the answer the test demanded.

The captain went through all the papers on the spot and called out our names.

'Right,' he said, 'you six are coming with me. The remainder stay here.'

Outside the room, we discovered our happy fate. We were going to train as VC10 pilots. The others would stay on as 707 navigators. It would be another two-and-a-half years before they would be called on to train as pilots.

After some basic classroom instruction on the workings of the VC10, the first part of our preparation for getting behind the controls of an *actual* VC10 would be via a series of sessions in the VC10 simulator, which, like all BOAC's simulators, was also housed at Cranebank.

The instruction module that would kick off our stint in the sim would be learning how to handle the aircraft at three very important points during take-off – the 'three Vs': 'decision speed' – the point at which during the take-off run when you had to commit, come what may, to take off, known as V_1; 'rotation speed' – the point at which you pulled back on the controls to lift the aircraft into the air – known as V_R; and the speed at which the aircraft could safely climb after take-off with one engine out of action – known as V_2.

I wasn't entirely sure how the following happened, but I suspect it had something to do with my having pontificated a little too loudly one day about the need for so much relentless drilling *by rote* of these procedures.

Right from the start of my flying – from before the day I'd gone solo with Big Bob in that tiny Cessna at Luton – I'd known to pay particular respect to certain points in the flight envelope. But did we really have to train in this way – having it rammed into us like we were kids?

Underpinning the question – and my frustration in the classroom on that day – was the arrogance, and perceived invincibility, of my youth – and a thought: these worst-case scenarios . . . *none of them is ever actually going to happen to me.*

The next day, walking into class before everyone else, I found a brown manila file on my desk with a report inside containing details – far too many, as it turned out – about another BOAC Boeing 707 crash, this one describing the final flight of BOAC flight 911 three years earlier – on 5th March 1966.

Speedbird 911 was on a round-the-world trip originating out of Heathrow until it broke up in mid-air over Mount Fuji, Japan. The aircraft took off from Tokyo Haneda Airport in the early afternoon of the accident, having taxied past the wreckage of a Canadian Pacific Douglas DC-8 – an aircraft that looked almost identical to the 707 – that had crashed in darkness and fog on approach to Tokyo-Haneda the night before, killing 64 of the 72 people on board. It is hard to imagine what the passengers of Speedbird 911 would have felt as they passed the burnt-out wreckage of the DC-8, but one thought that perhaps went through the minds of some of them as their 707 lifted into the clear blue *en route* to Hong Kong would have been: *As terrible as that sight is, statistically it reduces the odds of anything untoward happening on* this *flight.*

Prior to take-off, 911's captain, Bernard Dobson, sought and received permission from Tokyo air traffic to make a pass just to the east of Mount Fuji, where the air was as clear as it now was over Haneda. What Dobson didn't know was that a phenomenon known as a 'rotor' or 'standing wave' of highly turbulent swirling air found in the lee of tall, prominent mountains like Mount Fuji was active that day, brought on by unusual meteorological conditions. A succession of rotors, in fact, known as a 'mountain wave' and completely invisible to the naked eye, had formed in Fuji's shadow. Descending in a shallow dive from 16,000 feet towards the 12,388 feet peak of Japan's sacred mountain, the aircraft slammed without warning into the wave, exerting a massive lateral leftward force on the airframe. It came on so suddenly that it snapped off the aircraft's vertical stabiliser and sheared all four engines from the wing. The snapped fin broke off the left-hand horizontal stabiliser as it fell, causing the aircraft to pitch up. This caused the remaining empennage – the right-hand stabiliser and the rest of the tail section – to break away, sending the engineless, tailless aircraft into a flat spin. Horrified tourists below watched – and some filmed – as the aircraft spiralled like a falling sycamore seed, before scattering across Fuji's snow-covered slopes, killing all 124 people on board.

To begin with, no one knew why Speedbird 911 had crashed. Clear air turbulence of such violence was, at that time, an unknown phenomenon. It was only when investigators found and analysed the intact 8 mm film on the cine-camera of one of the passengers that the truth began to dawn on them.

The film depicted the last moments of all on board Speedbird 911: from the wreckage of the DC-8 that they had taxied past on the way to Haneda's threshold; to shots of Mount Fuji as Captain Dobson had put the 707 into that fateful shallow dive over the mountain.

When the investigators found two skipped frames on the developed film, they realised this was the moment the aircraft had begun to break up. Subjecting the camera to tests, they also realised it required an immediate onset force of 7.5g – seven-and-a-half-times the gravitational force we experience on *terra firma* – to have caused the frames to skip in this way.

From there, everything else fell into place – the mountain wave phenomenon was identified as the culprit and, happily, not a single airliner has been lost since to the same phenomenon – a testament to the adage that the science of air accident investigation advances one crash at a time.

It is also a testament, sadly, to the fact that the whole science of air safety does too.

As I closed the file, embarrassed that it had taken my snap, ill-chosen comment to prompt an instructor to have left it on my desk, I took on board a number of things.

First, respect the process. Twenty-five years after the end of the Second World War, for all the wonders of our new age – modern jet airliners like the 707, the DC-8 and the VC10 – aviation was still far from being a mature technology; like air safety, it also advanced a generation at a time.

Second, the things that you can't or shouldn't do on a flight deck – or anywhere else around an aircraft – are built on foundations of understanding.

But for Captain Dobson's fateful decision to have deviated from his schedule, 124 people who died that day would still be alive. What had led him to do this? The file offered no answers to this question, but as we approached the next phase of our training, a voice whispered in my ear: pay attention – pay attention to every little detail – because one day it might save your life.

Had that same voice whispered in the ears of the Air France crew that had crashed at Gonesse during *their* training?

This, too, is impossible to know, but the evidence said, at the very least, that some fundamental lessons learnt had been forgotten or ignored on that terrible day.

4

'Engine fire Number 4!'

The shrieking, piercing bell got my attention the second it went off. The Number 4 engine had burst into flames as we were approaching rotation speed, V_R.

Concentrate, Mike, concentrate. Thoughts of Speedbird 712 *Whiskey Echo*, flooded my thinking. I snapped myself back to the present. *Whatever you do, don't get distracted by it. Rely on your training . . .*

I rotated, keeping the aircraft straight and, as soon as we were climbing, called 'Gear Up'. I had to be spot-on now that we only had three engines running properly – holding to our V_2 minimum climb-out speed, which required me to maintain a three-degree climb-out gradient after clearing 35 feet at the end of the runway.

As soon as the non-handling pilot, the captain, and I had agreed the flight engineer's diagnosis was right, we went into the drill.

I cancelled the ear-piercing bell and called out: *'Fire-action Number 4 engine!'* – trying my damnedest to sound calm and professional.

This triggered us all to go straight into the checklist of procedures, for which, up to this point, we had drilled relentlessly. The immediate actions needed were all from memory.

As I was physically flying the plane, most of the actions were being carried out by the captain, acting as my co-pilot and the flight engineer. *'Fire-action Number 4 engine,'* they repeated in unison, to confirm we were working together as a team to put the fire out and keep this 150-tonne beast under control. Whilst I flew the aircraft, they'd announce what they were going to do – and, when we were all clear and had agreed the actions, they'd begin to initiate them.

On the coaming were four T-shaped fire handles, one for each engine. In the handle, there was a red light. It told an alarmingly simple story. If it was on, and it was, that bloody engine was on fire.

With the bell silenced, the next action was to close off the fuel to that engine by shutting its high-pressure fuel cock. That was a job for the flight engineer. But because he was shutting off the fuel supply, it was imperative he moved the correct switch. Should he inadvertently stop the fuel supply to the Number 3 Engine, we'd have an even bigger problem: instead of one engine, we'd have two engines out – and an aircraft I was trying to fly at maximum weight would plummet like a stone.

Moving steadily, and speaking clearly, the engineer said: 'Number 4 HP cock – shut.' There was a pause between 'HP cock' and 'shut' to allow us to look and confirm he had the correct switch.

Next: pull the 'Fire Control Handle' for the Number 4 engine. That should automatically do a number of things in our bid to stop the fire.

The red light was still on in the handle, so, though it was obvious which one to pull, we had to do it by the book; stick to those standard operating procedures, those SOPs, that had been drilled into us.

So, the captain carefully, and deliberately, placed his hand behind the handle, fingers around the stem, and said: 'Engine Fire Control Handle — pulled'.

Again, a pause so we could check that he had the correct handle before he pulled it.

This automatically closed the other, low-pressure valve supplying fuel to the engine. At least, that was what it was supposed to do. The flight engineer confirmed that it had worked properly by saying: '*Number 4 LP cock checked shut*'. If it hadn't worked, he'd have closed the low-pressure fuel cock manually.

I hoped that this action would kill the fire. To be doubly sure, as soon as the fuel was shut off, we looked to see if the red light was still there. If it was, the captain would fire an extinguisher shot into the engine.

It was – so we did.

On the VC10, the fire extinguisher controls were built into the fire handle. To activate the first extinguisher, you twisted the handle clockwise. The captain said, 'Number 4 Engine – first shot fired.'

Again, a pause to check.

Whilst we waited to see if the light would now go out, the flight engineer did his 'clean-up' drill, checking that the hydraulics, low- and high-pressure fuel supplies had shut. Just to be doubly sure, he closed the switches manually anyway. That done, he reported: 'Immediate Action completed and checked.'

All eyes back on that fire light.

Still on . . .

A knot formed in my stomach. My mouth went dry.

Give it thirty seconds . . .

The wait seemed interminable.

I looked back at the light. Had it gone out?

No, damn it . . .

So, the captain fired the second extinguisher by turning the handle anti-clockwise. Meanwhile, I focused with all my might on continuing to fly at the right climb-out angle and speed, the V_2 that we'd calculated before take-off.

I glanced back at the light and breathed a sigh of relief. Finally, it was out.

With the aircraft under control, I told air traffic control what we were doing, prefacing it with that classic call: 'Mayday, Mayday . . .'

Even though I was now in control of the aeroplane, my problems weren't over. I now called for the 'secondary actions' to be carried out.

On the flight deck, you had a manual of 'normal checks' – the kind you go through before engine-start, after start and after take-off – but in an emergency, you had to refer to another manual, the *emergency* checklist.

The flight engineer grabbed the list and read it out to confirm that all of the immediate, memory actions had been correctly completed.

When he was satisfied, he called: 'Engine fire drill completed from the checklist.'

He had now ensured everything we'd just done had been done by the book.

A beat. Then . . .

'That's great, Mike,' said the instructor, leaning forward from the seat behind us, cool as a cucumber. I realised, meanwhile, that I was sweat-soaked. 'Now, let's go and do a three-engine approach in limiting weather conditions – and a crosswind.'

He pressed a couple of buttons and the simulator instantly repositioned to 10 miles out onto final approach at Heathrow, just like magic. And, in a way, that's exactly what a simulator was . . .

*

The VC10 sim was a computerised, valve-driven behemoth hooked to a single-axis motion platform and a visuals system that simulated your route by projecting images of a model landscape – a 3D model of a generic runway made to represent any runway, anywhere in the world, that some poor so-and-so had had to go and build – onto a TV screen in front of the pilots.

To get to the simulator, I'd already been through a four-week technical course in a classroom with an exam at the end of it. The course had provided basic instruction on the VC10's flight characteristics, its Rolls-Royce Conway jet engines, how the radios worked and much other technical stuff. After this, I'd had a week of instruction on its safety features: the life-rafts, emergency exits and chutes, fire extinguishers, how to use the survival kit, *et cetera*.

The simulator was my first experience of 'handling' a big jet – an opportunity to learn procedure before I'd ever set foot in an actual VC10.

All flying emergencies, like the one I'd just been through, had to be practised and mastered – from the kind you were supposed to get checked on every six months, such as an engine failure on take-off and engine-outs on approach and landing, to checks that were more 'routine' – emergency depressurisation drills, for example – on which you were evaluated over more extended cycles, every three years or so.

With so much to learn in such a short time, it was like trying to drink water from a pressurised fire hose.

The big ones – an engine failure on take-off, a subsequent engine-out climb away, an engine-out on approach and landing (and go-around, if, for any reason, you couldn't get into the airfield) – were things you were drilled on, and regularly checked on, because they were the most critical potential failures.

A Rolls-Royce Conway is a big turbofan, has lots of fuel running through it, pushes out a lot of power and – as happened on Flight 712 *Whiskey Echo* – can, on ultra-rare occasions, fail catastrophically.

An engine failure can also significantly affect the handling of an aircraft and some of the systems that are driven by the engine will stop working.

The VC10 had four Conways mounted in a cluster at the back. The clustering of the engines had led to the most marked feature of the VC10 – its high, swept tail. In order to keep the tail surfaces out of the way of the jet exhaust, the horizontal stabilisers had had to be mounted high. Funny, I thought, that such a hard-edged engineering decision had led to such a pleasing aesthetic outcome – the grace and beauty of the VC10 itself.

Of my ten four-hour sessions on the simulator, a large part of them hinged on training for the three Vs.

Take-Off with an Engine Failure

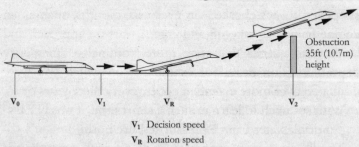

V_1 Decision speed
V_R Rotation speed
V_2 Climb speed after engine failure

V_I was the first. For every take-off, you needed to calculate the speed at which you could safely stop on the runway, using the brakes and reverse-thrust on the engines that hadn't failed.

Before engine-start, you knew the aircraft's weight, the length of the runway, the outside air temperature and the prevailing wind conditions. These would allow you to calculate how much runway was needed for take-off. Most runways are long enough not to restrict you by weight – this was especially true of the VC10, which had been designed to take off from short runways.

Your initial calculations are made on the basis all four engines are working. But if an engine failed, even on a longer runway, you were what's known as 'limited'.

Critical to this calculation were two positions: one of them denoted a point on the runway beyond which, if you tried to stop, you wouldn't have enough concrete left to be able to do so safely; the other – which came before it – was the spot at which you still had enough speed to get airborne.

Between the two points was a band that said you had enough runway either to stop or get airborne in the event of an engine failure on take-off.

Trusting yourself on the spot to make that decision – if or when it ever happened – was a complete no-no, no matter what aircraft you were flying. Human nature being what it is, most of us would dither as we tried to weigh our options.

A point, thus, was picked beforehand – a 'decision-point' – at which you either slammed on the brakes and applied reverse-thrust or committed to take off.

In reality, you'd probably be able to stop safely beyond this; and to get airborne before it. But the decision-point forced you to focus.

In a VC10, on a regular runway, with a normal take-off weight, average outside air temperatures and prevailing winds – i.e. under so-called 'non-limiting conditions' – you might be looking at 150 knots as your safe stop-point and 110 knots as the first point at which you might safely get airborne, with your decision-point falling between the two – i.e. around the 130-knot mark.

If the runway was more 'limiting', however – i.e. if it was shorter than you'd like, at a higher altitude, if the outside air temperature was higher, or the prevailing wind speed slower, or coming from behind you – the band between the two markers closed to a point where there might not be a gap at all.

The situation you didn't ever want to find yourself in – all this in the rare event that you'd lost an engine on take-off – were conditions in which the two markers crossed over: at a point, in other words, where the first opportunity to get airborne fell *after* the very last moment you were safely able to stop.

The decision-point, known as V_1, is drilled into you right from the start of your flying career, but is only really critical when you're flying a big, heavily laden jet.

An engine failure *after* V_1, the exercise I had just been through in the sim, was altogether different.

As you barrel towards the point at which you rotate – the moment you ease back on the stick, known as V_R – you've got the same issue of keeping the aircraft straight, only your rate of acceleration has decreased because you've lost an engine. The moment you get into the air, the aircraft wants to yaw into the bad engine, but you've got to keep climbing on a track that was determined before take-off, because there may be hazards close to the airport – mountains or tall buildings.

Once you've got the aircraft under control, you bring the undercarriage up to reduce the drag. You don't leap into your emergency drills because you need to do everything by the book, and anything you *do* do is with the agreement of the crew.

In the sim, we were taught metaphorically to sit on our hands until the aircraft got to 400 feet – and you're there almost before you know it.

On the VC10, there were three crew on the flight deck – two pilots and a flight engineer.

This may seem blindingly obvious, but the first thing you had to do was to agree amongst you what had gone wrong.

This is where you needed to be methodical – you have an opportunity now you're in the air to analyse the nature of the problem. You've had an engine fail – but *why* did it fail? Is it on fire? Is it a catastrophic failure like *Whiskey Echo*'s?

The chances are it'll be the flight engineer who identifies the problem.

As a pilot, you've felt the aircraft lurch right into the bad engine at the moment of failure, but with four engines at the back, you don't automatically know which of the two on the right-hand side has failed because you're fully occupied, looking up and out of the cockpit. That's why it's left to the flight engineer.

If it's a fire, there'll be a warning light you can see and a bell you can hear – but it's still left to the flight engineer to diagnose the problem. Then, between the three of you, you need to agree that his diagnosis is the right one. This means first agreeing that the engine he's called out as having failed or caught fire is correct; and second, that the problem he's identified – a failure or a fire – is correct also.

It was for this reason that we had SOPs – those standard operating procedures – in which we were trained to do certain things and expected others to do certain things at certain times as well.

After several sessions on the sim, I had mastered the VC10's main handling characteristics under the most critical emergency conditions.

Next up was flying the aircraft for real – a part of the training regime that would take my fellow cadet pilots and me to Shannon in southern Ireland.

There, we would be tested on almost all the drills we'd carried out on the simulator – including a couple, when I thought about them, that had given me sleepless nights: in particular, recovering from what are known as 'Dutch rolls'.

Pilots I'd spoken to who'd been through this had told me I needed to prepare for the fact I'd be flying the VC10 more like a fighter than an airliner.

5

Heathrow, on a clear morning in early September, a few days after the final segment of my simulator training, and I was standing beneath the plane known as the 'Queen of the Skies', gazing in awe at her most distinguishing feature, her swept T-tail, which seemed almost to scrape the high cirrus.

The sunlight glinted on the polished white and blue surface of her fuselage. My gaze moved from the four Conways, clustered purposefully between the tail and the ground, to the golden 'speedbird' on her blue fin.

Today was the day we were flying to Shannon to practise some of the more extreme manoeuvres you couldn't do at Heathrow, including 'upper air work' and rejected take-offs. I would take the right-hand seat on the flight out.

Despite its order of fifteen 707s in 1956, BOAC had gone on to buy the British VC10 because of an inherent weakness in the Boeing-designed jet – an initial lack of power in its Pratt and Whitney engines. This deficiency meant that BOAC would not be able to fly to certain 'hot and high airports' it was committed to serve. Indeed, the 707's performance out of Addis Ababa, Nairobi and Mexico City was so limited that the airline's only option would have been to reduce the fuel and number of passengers it carried on those routes.

To get around the problem, BOAC had ordered the VC10, a beautiful design with some real innovations. 'Fowler flaps' along most of the trailing edge and leading-edge slats covering almost the full span of the front of the wing helped – along with the 22,500 pounds of thrust produced by each engine – to give the aircraft its exceptional short-field, hot-and-high performance.

These features, along with her sturdy undercarriage (specifically for those rough, short fields in Africa and South America), clean fuselage and elegant swept wings, were all in evidence in my walk-around on that crisp, early autumn morning.

My instructor pilot, Geoff Morrell, in the left-hand seat, was one of four instructor pilots who would be with us in Shannon. Sitting in the cabin were three other pilot instructors, four qualified pilots (needed because you had to have two fully qualified pilots on the flight deck at any one time), my seven fellow cadet pilots, a couple of instructor flight engineers and a handful of ground crew, who would support the VC10 on her two-week deployment.

A veteran flight engineer instructor, Arthur Winstanley, was in the flight engineer's seat behind me.

There were two types of instructors: the 'soft cuddly kind' (as we cadet pilots mistakenly thought of them), who included Morrell and Winstanley, and the more gruff, taciturn variety.

The latter included Bob Knights, known affectionately as 'Chopper', who, right now, was sitting with the others in the back of the aircraft. All of them had flown bombers – Lancasters, Halifaxes, Wellingtons and other types – during the war. Bob Knights was ex-RAF 617 Squadron, the Dambusters, and had been one of those who had attacked the German battleship *Tirpitz*.

As I taxied out, besides the practice I'd done on the sim, the only point of reference I had for steering an aircraft on the ground was driving a car.

On a car, steering is intuitive – you turn the wheel, you touch the brakes, and the car does what you want it to. Not so with a big airliner.

For a start, the wheels aren't level with your feet, as they are in a car, but some way behind the cockpit. And on the VC10 there was a strange device called a 'tiller', a handle to the right of my seat, below the window, which was used to steer the nosewheel, giving greater control than the pedals, which only directed the travel of the wheel so far. The tiller was far from instinctive – you pushed it forward to turn left and pulled back to turn right. Control was normally in the captain's hands, but the tiller on my side was linked to his by a cable. I worked it gingerly as we headed out towards Runway 28 Left.* Unable to see the wingtips, I had to think some way ahead. Did I have enough clearance between me and that obstacle? If I went down that taxiway, could I get out of it?

It was, in hindsight, an enormous act of faith on BOAC's part – no matter how much sim time we'd had – that they allowed us to transition straight from Hamble's piston-driven twins to a jet the size of the VC10. It doesn't happen today. The only route to a big jet in British Airways – a Boeing 777, say – is first to have put your hours in on a short-haul jet like the 737.

A final go-through of the take-off checks as I turned on to the runway, then: 'Speedbird Three-Four Golf, clear take-off Two-Eight Left, wind two-one-five degrees, seven knots.'

Geoff, as the non-handling pilot, dipped the transmit button: 'Clear take-off Two-Eight Left, Speedbird Three-Four Golf.'

* It changed to 27L in 1987 as magnetic variation had reduced.

A rill of excitement in my veins as I pushed the thrust levers forward. The Conways answered with a noticeable thump in my back and the runway started to roll. After what passed in a moment, Geoff called 80 knots.

We hit V_1, then, moments later, V_R. I gently pulled back on the stick and the Queen of the Skies roared into her domain. Our climb-out looked good. I commanded 'gear up'.

Moments later, we were cleared by air traffic to 6,000 feet. I glanced down. As the sunlight glinted off the Wraysbury Reservoir, I brought us on to a heading of two-eight-five degrees and a course for southern Ireland.

6

The weather at Shannon could be shocking – and that September we were treated to autumn gales that kept us on the ground for much of the time we were meant to be in the air. The advantage of having some of the hoary old instructors with us (in actual fact, they were mostly in their forties, but being former bomber crew, they seemed old and hoary to us) was that they refused to let a simple thing like the weather stand in the way of our education – and if we couldn't get into the air, they found something else for us to do. In Ireland, that meant a trip to the pub or a round of golf at Kilkenny or Lahinch.

Geoff Morrell loved his golf, but the first day he took me to Lahinch it was blowing such a gale that when I teed off, the ball lofted into the air, got caught by a gust and ended up 20 feet behind me. We abandoned the game and went to the pub instead.

The next day, the weather had cleared enough to do some upper air work.

All pilots are taught *ab initio* how to recover from a stall. On a Chipmunk, it's simple. As your airspeed drops and the aircraft begins to depart from controlled flight, you push the nose forward, the speed builds over the control surfaces and you recover.

On a VC10, it was altogether different. The aircraft's most graceful feature – that high T-tail – had the capacity to be a killer. Six years previously, another aircraft type with the same aerodynamic configuration – a BAC 1-11 prototype conducting stall tests over southern England – crashed into Cratt Hill in Wiltshire. This was the first crash attributed to a phenomenon known as 'deep-stall' – also known as 'super-stall' – which occurred when the turbulent wake of the stalled wing blanketed the control surfaces of the T-tail. In such a condition, the elevators become inoperative, whatever you do to the controls is ineffective and the aircraft 'mushes' with next to no forward airspeed into the ground. This is what happened to the 1-11 and within three more years, deep-stall would claim a second British T-tail victim, a BEA Trident operating out of Heathrow, which crashed a short distance south of the George VI reservoir near Staines, minutes after take-off, killing all on board.

To overcome the deep-stall on T-tail aircraft with rear-mounted engines, a device called a 'stick-shaker' was introduced. This, as it sounds, produces a shaking movement in the control column as the aircraft's systems detect the onset of a stall. That was the moment you pushed the stick forward. If for any reason you neglected to (there were no good reasons why you would), a secondary device called a 'stick-pusher' automatically did it for you.

The person who took me through this was Chopper Knights who, though more friendly than he appeared, liked to play up to his reputation.

Chopper was sullen and wiry, and older than the other instructors, but probably still only in his late forties. Being an ex-617 Squadron Dambuster, arguably the most famous squadron of the entire war, Chopper had more than done his bit – and it was this experience that had launched him on his mission, which was to get his cadets proficient on instruments as early as possible.

Instrument training – on the simulator at Cranebank or in the real aircraft – involved a visual take-off prior to switching to IFR (Instrument Flight Rules). In the real aircraft, the instructor placed screens over the windows so you couldn't see out. You'd fly a track to a particular beacon, turn around, come back, fly a holding pattern, then an approach using the instrument landing system before going round and doing it again. Finally, the screens were removed to enable you to do a visual landing.

Chopper's mission was to get his cadets to do this very early on – more often than not, during the cadet's second flight – because, he told us, so few of his contemporaries – other wartime bomber pilots – had been proficient on instruments. Those that had been lucky enough to survive, I realised, had spent most of their war looking up and out for flak and fighters. And, heaven forbid, if they did need to navigate anywhere, they'd ask their co-pilot or, worst-case, their navigator which way to go.

It was a testament to the training we'd received at Hamble that when Chopper sprang this on us, all of us were able to do it without a problem.

So, when it came to my deep-stall test, I knew I was in good hands. The point of the exercise, Chopper reminded me, was to take it up to the stick-shake, not the stick-push.

After a couple of goes, I got the hang of it. The aircraft slowed, the nose came up, the stick shook, I pushed the nose forward, got a grunt of approval from Chopper, and on we went.

Geoff Morrell had an altogether different party piece. Roughly 30 miles north-west of Shannon were the spectacular 700-foot-high Cliffs of Moher – black and vertical, they emerged primordially out of the Atlantic as a potent marker for our run-in to Shannon whenever we returned from our upper air work training.

The first time I flew with Geoff, him in the left-hand seat, me in the right, he announced he had control and let down through the clouds and the spray until we were down to just a few hundred feet. In the distance, through a curtain of mist and spume, the Cliffs suddenly appeared. We were heading for them at a couple of hundred knots.

Scarcely above the waves, we flew closer and closer, until they appeared to be towering above us.

There was no way on God's Earth I told myself, as my fingers clenched the seat, we would be able to clear them. But at the last second, Geoff poured on the power, pulled back the stick and the VC10 shot over the top, clearing the cliffs by what seemed to be a few feet. Out of the corner of my eye, I caught sight of an isolated pub. Then we climbed away to join the circuit for Shannon.

Two nights later, we drove out to the very same pub, which proved to be a mistake. The landlord cottoned on that we were the BOAC crew that had been beating up the vicinity in a VC10 and complained to us vociferously, saying we should all be shot. He then served us a meal that would have been rejected by the condemned. Geoff Morrell seemed to take all this in his stride.

The next day, however, following some more upper air work, I found myself once more in the right-hand seat and Geoff in the left, letting down through the clag towards the Cliffs of Moher, pulling up and over them – only this time, Geoff veered straight for the pub. He took the VC10 down to 150 feet and flew bang over it, pouring on the power as we roared over the top. By the time we landed, the crusty old publican had already complained – and would, he said,

be sending BOAC the bill for four new windows. Geoff had
had the distinction of being one of the youngest Lancaster
pilots at the end of the Second World War. You didn't mess
with these people.

7

The VC10 was susceptible to Dutch roll because of the nature of its design – that beautiful clean swept wing, which gave it its grace and speed. An RAF VC10 set a record for the fastest crossing of the Atlantic by a subsonic jet airliner that held for 41 years, beaten only by a Boeing 747 a handful of years ago.

To counter this susceptibility, Vickers had added what are known as 'yaw-dampers' – accelerometers and sensors that monitor an aircraft's rate of yaw, sending signals to the control system that automatically controlled the actuators on the rudder, killing the aircraft's tendency to roll in certain corners of the flight envelope.

The VC10 had an early version of an auto-stabilisation suite that is now standard across all airliner types. But we had to demonstrate an ability to spot early onset Dutch roll and recover from it, an exercise that involved switching off the dampers and inducing yaw by putting in a great big boot of left rudder.

Simplistically, as the aircraft begins to yaw left, one wing travels marginally faster than the other and begins to develop slightly more lift, causing the onset of the roll. As it rolls, both wings develop equal lift and the nose drops. Because of the aircraft's big fin, it then yaws the other way. If you do nothing, the aircraft becomes 'divergent' – this is where the amplitude, the maximum extent of the movement, becomes greater and greater until, if you continue to do nothing, it'll flip onto its back.

The unnerving thing was that you had to let the aircraft build up a lot of bank before you got involved. The instinctive thing to do was to put control in to lower the right wing early to stop it happening. But that made the right wing react too quickly and increased the amplitude. What you had to do instead was wait until the aircraft passed through level, rolling right hand down, and then put in the control you know it's *going* to need shortly – a load of left-hand aileron. When it gets back through level the other way, you put in right-hand aileron, which dampens it all down.

The exercise was nauseating and not just for those in the back – it could also make the most hardened stomach heave if you were in the pilot's seat.

For the Dutch roll test that I undertook, there were two safety pilots on board – one of whom was a charming, if slightly dour, Scot called Bill Ferguson. Bill was a little old for a co-pilot, but still only in his early forties.

Bill had been on a bender the night before with some of the others and wasn't anticipating he'd be flying first thing. Fortunately, due to other safety pilots being on board, he didn't have to sit up front for my Dutch roll test.

As I got up from my seat and walked forward through the cabin *en route* to the flight deck, suppressing a sickening feeling in my stomach, I passed Bill stretched out across three passenger seats, fast asleep.

God, I thought, *how can he do that at a time like this?*

There were four cadet pilots on board. One of them was up before me and already in the right-hand seat. I was up next.

As I stepped on to the flight deck, I was trying hard to keep all the details of the Dutch roll briefing in my mind. But with the stakes high, and feeling the tension, I'd seemed to have forgotten most of it. There was very little room for misunderstanding or control input error on a Dutch roll test. I also knew – because you could see it on their faces – that it was the exercise the instructors were most tense about too. Dutch roll testing was as close to the flight envelope boundaries of a civilian airliner as you'd ever want to get.

I stood at the back, watching the cadet pilot before me like a hawk. We're taught from an early stage as pilots to think ahead. If such and such were to happen in the next second or two, what would I do? After a while, this forward-scanning mode becomes instinctive – like that thing we do when we're driving: before we know it, we've gone 5 miles on a route we know well without the least memory of how we got there. Soon, the same instinct would take over when I was flying – a part of my subconscious would always be scanning ahead for trouble. But not today. Today was about drilling for something that was never expected to happen but would be catastrophic if it did.

My cadet pilot colleague had already been through a few practice runs; the VC10 had been corkscrewing through the sky, and queasiness, accentuated by my nervousness, was starting to get the better of me.

When his turn was over, my colleague clambered out of the right-hand seat, and Geoff Morrell motioned for me to take his place.

Looking back, I was the most nervous I'd ever felt while flying a VC10 – a lot more so than on the deep-stall test. Then, I'd known that the aircraft was protected by the stick-pusher. And the captain would have intervened if, for a second, he'd thought we were in danger. But here, if I put the wrong control inputs in at the wrong time, the VC10 would be on its back quicker than the captain could intervene – something I *had* recalled with great clarity from the briefing. The only time I ever felt more nervous was a test-flight once on Concorde.

'OK, Mike?' Geoff said, as I strapped in.

'Yes,' I replied, doing my best to smile.

'All right. I'm switching the yaw-dampers off and I'm going to put in a large amount of left rudder to generate the forces we're looking for. Got it?'

'Got it.'

'You're going to rest your feet on the pedals and follow my actions.'

'OK.'

'When the aircraft starts to diverge, we leave it. To begin with, the roll will be barely perceptible. But because it's divergent, it will increase. Ready?'

'Ready.'

Geoff put his boot into the rudder and the plane started to crab. The horizon slewed sickeningly and the right wing started to come up. Feeling the movement through the control column, I thought Geoff had to be responsible for the inputs. I glanced left. He wasn't. We were pirouetting like a tossed cork on a storm swell. Nightmarishly, like a bad movie, the VC10 was flying itself.

As the wing came back down towards the horizontal, Geoff said: 'When we go through wings level, I'm going to put some counter-aileron in. Keep your hands on the control column and feel the amount of input that I give it.'

I did. It was liberal, but smooth.

'Now, I'm going to let go,' he said. He waved both his hands to show me they were free of the control column. The aircraft was still going through its crazy, mind-of-its-own routine, but I could feel it starting to self-correct.

'It's intuitive that when the aircraft begins to roll back again that you'll want to put some aileron in. *Don't*. Let it sort itself out because the only way the Dutch roll will dampen down is of its own accord. If it doesn't, that's when you need to put aileron in – into the next cycle. You got that, Mike?'

'Yes,' I said.

'Good. Because now *you're* going to do it.'

The queasiness I'd felt on my long walk up to the flight deck intensified.

To induce the roll, I did as he had done, only in the opposite direction – I gave a big boot of right rudder and the aircraft started to move. Geoff was correct. I felt a huge desire to counteract with left rudder but held off. The urge built as the left wing rose. Then, all on its own, the VC10 started to roll the other way. Don't do anything. The point of the exercise was to let the forces build.

And they did. Before I knew it, the right wing had gone past 45 degrees and was building towards the vertical. I glanced out the window and saw the white tops of the waves far below and had a momentary vision of the aircraft on its back. A voice in my head began to yell – *when do I take action?* All on its own, the aircraft rolled the other way. As we went through horizontal again, Geoff said: 'Corrective action, *now*!' I gave the aircraft left aileron. The VC10 continued to roll towards the right, but I felt a slight dampening in the forces through the controls. And as the wing barrelled down the other way and then came back up, I gave it some aileron the other way – a little less this time. The motion, just as Geoff had said it would, slowly began to correct itself and eventually stopped altogether. Within moments, we were flying along as if nothing had happened.

'And that, my boy, is how you get yourself out of Dutch roll,' Geoff said.

We practised it two more times until I'd got the complete hang of it.

Then I got out of my seat to let the next guy have a go.

He was standing, as I had been, at the back of the flight deck, hanging on for dear life. Seeing the greyness of his complexion, my sickness returned and I hastened past him to the toilet in the cabin – regretting by now the large breakfast I'd enjoyed a few hours earlier at the hotel. Before I'd got halfway, Geoff started the routine all over again as he demonstrated what he'd done for me to the next cadet pilot.

As the aircraft began its sick-inducing yaw-roll motion, the curtains halfway down the cabin began to swing. To stop myself from throwing up on the spot, I tried to tell myself that the curtains weren't swinging; it was the aircraft that was pirouetting around *them*. But that only made it worse. I got to the toilet just in time, but then a minor miracle: one whiff of its blue Elsan had a curiously restorative effect on me. There was no way I was going to throw up, I told myself. I'd just mastered Dutch roll.

I moved back towards my seat in the cabin. On the way, I passed Bill Ferguson, who was still curled up like a baby where I'd last seen him – across the three seats. I could still hear his snores when I sat down half the cabin away.

Dutch roll and deep-stall are two hazards that have been eliminated as risks from passenger flying thanks to technology. A third, to which the VC10 might also have been prone, was a nasty phenomenon known as 'Mach tuck'.

Because it had lots of power and was aerodynamically clean, the VC10 was, quite extraordinarily, probably capable of going supersonic – something, because it was never

designed to fly at such speeds, that would have been fatal.
Mach tuck is associated with swept wing designs at high altitude and occurs when the centre of lift moves aft on the
wing, pitching the nose down. The more the nose dips, the
faster the aircraft goes and the further aft the centre of lift
moves. Shock waves on the wing reduce the effectiveness of
the controls and, like Dutch roll and deep-stall, without
intervention would cause the aircraft eventually to break
apart.

However, as with the other two phenomena for which we
trained, even in the sixties the automated systems designed
into aircraft like the VC10 were so good – so technically
advanced – they prevented this from ever happening.

Being the last test of our Irish sojourn, we packed up late
and headed back in the twilight towards our hotel in Ennis –
in a rental car driven by Arthur Winstanley, the veteran flight
engineer. Geoff Morrell was in the passenger seat. I was in
the back. For some reason (we didn't like to ask lest they let
loose with a judgement about our flying), we whippersnap-
pers were always kept well away from the wheel of a car. It
was only ever the instructors who drove.

As we approached Ennis, Arthur was chuntering along
happily when a Garda officer jumped out from behind a
bush and flagged us down. The officer walked leisurely up to
the car. Arthur wound down his window.

'Good evening, *Sor*,' the officer said. He was a ruddy-faced
man of around the same vintage as Arthur and Geoff. 'Do
you know what speed you were doing there?'

'About thirty?' Arthur replied, hopefully.

'No, it was a tad more than that, *Sor*,' the Garda man said,
shaking his head. 'I'd reckon you were doing around fifty.'

'Oh,' Arthur said, feigning mild surprise, 'well, then, I suppose I must have been.'

'What, then, *Sor*, do you think I ought to be doing about that?' the Garda's finest enquired of him.

Arthur thought for several moments and said: 'Why don't you stand further inside the 30-mph limit so people can slow down when they see you?'

Up until this moment, the exchange had been of such a good nature that I'd gone along with it, nodding and smiling with Arthur and the Garda officer at all the right moments. But an ominous look seemed to cross the policeman's face for a second or two. He took a deep breath. We waited, then he said: 'Actually, *Sor*, that's a jolly good idea. I'll do that.' And off he walked.

Feeling happy that we'd got off so lightly – and, no doubt, feeling the need for a little fortification after a day of aerobatics in a fully-fledged airliner – Geoff and Arthur expressed a desire for a pint of Guinness soon after we parked up behind the hotel. The hotel looked dead, so we ambled off down the street looking for somewhere that would serve us a pint of the Dublin Mudslide. But everything seemed very shut. I looked at my watch. It was not even 10 o'clock.

Eventually, we came to a tavern from which an encouraging glow shone behind the glazed windows.

I tried the door, but it was locked.

Not to be deterred, Geoff marched up and banged on it loudly. As he stepped back, we heard a voice in the shadows: 'What is it that you're doing there, *Sor*?'

We turned as the Garda officer stepped into the light.

'Ah, good evening again officer,' Geoff said, before adding sheepishly: 'We were hoping for a bit of refreshment.'

The officer stared at us, hands on hips for a second, then

marched up to the door and banged on it so loudly that I thought it might fall off its hinges.

'O'Malley!' he yelled. 'O'Malley! Open up!'

A sliver of light appeared at an upper window. Then a face poked through the curtains. My imagination after all these years may have provided Mr O'Malley with a candle; he was, though, very definitely wearing a bedcap.

'What is it that you wanted, Padraig?' he shouted down to the officer.

'It's a BOAC crew and they're wanting a drink!' he yelled back.

Two minutes later, there was a rattle of chains and a tumbling of locks. We stayed drinking with the hapless O'Malley and Padraig the Garda officer until 2 a.m. As we were flying back the next day, we finally agreed between us that it was probably time for bed.

8

I came back from Ireland fully qualified to fly the VC10. A few days later, I flew on my first trip – to Nairobi.

The VC10 operated with a captain, two co-pilots and a flight engineer. Though qualified to fly the aircraft, I was not the second-in-command and wouldn't be for a while – that honour was bestowed on the senior first officer, known internally as a 'P2'. Still under training, I was a lowly P3, a second officer with one stripe, and loving every minute of it.

The big difference between a P3 and a P2 was that the latter was also qualified as a navigator.

Hard to believe today, but the VC10, like the 707, needed to operate with a navigator on board for the trans-oceanic and trans-Sahara legs that we undertook, during which our main navigational aids were the Sun, the Moon, the stars and Venus.

Some days, under the captain's instruction, I'd do the take-offs and landings; others, the P2 would. If the latter, he'd then slide out of the right-hand seat, I'd jump in and fly the rest of the sector, while he did the navigating.

To become a P2, I had to become a fully qualified navigator, which was somewhat ironic as I thought I had dodged that bullet all those months ago.

I now put my name down for the navigator's course.

Becoming a P2 was kind of a numbers game – but it required, too, an ability to show enthusiasm and to duck-and-weave. To become a P2, there was no dodging the navigator's course; *when* you did it, though, was up to you.

There was, in fact, some incentive not to do it too soon, because in the early seventies a system came into BOAC called 'bid-line'. Up until then, the trips we took were all rostered. I'd get a message saying: 'Right, Mr Bannister, you're going to Nairobi next Thursday.' And that was it.

But then the union stepped in. It wanted a system where the pilots had some say over when they worked and where they went.

After the union's intervention, all the work that needed to be done was published in 'lines of work' during a particular calendar month, and you'd bid for the work that you wanted to do – hence the name.

The more senior you were, the more likely you were to get what you wanted.

You were only allowed to bid within your category: captains had one set of bid-line; senior pilots/P2s another; and junior pilots/P3s a third. You bid, in other words, amongst your colleagues, who were at the same level as you.

But then came the duck-and-weave bit. By doing the navigator's course and becoming a P2, you'd go to the bottom of the P2 pack. If you didn't do the course and remained a P3, you'd rise gradually to become the top of the P3s.

In terms of control over your social life, destinations, your days off and, to an extent, the money you earned – certain trips earned more in basic pay and allowances than others – you might actually be better off staying on as a junior pilot.

On the other hand, if you were ambitious – and I was – to increase your qualifications, training and experience, you'd say, to hell with that, and go for a higher position, knowing that, in the shorter term, at least, you'd be worse off financially.

While I was weighing up what to do, I found myself flying trips to Mauritius, the Indian Ocean island 600 miles off the coast of Madagascar, which BOAC (and later, British Airways) served via Nairobi. Not that I knew it at the time, but this idyllic paradise would later act, too, as a formative link in a chain of deduction that would end in a Paris court after the Concorde crash.

Mauritius was a destination that took crews away for eight days. You flew out from Heathrow on a Thursday and flew back in on a Thursday. Not surprisingly, it became a regular bid slot for me – eight days in which there was nothing much to do except lie on a beach, wait for the turnaround – and get paid for it.

During the several occasions I ended up there in the late sixties and early seventies, my colleagues and I got to know a Qantas 707 crew quite well. This lot would fly to Mauritius out of Perth in Western Australia and often overlap with us. Given the (largely) good-natured leg-pulling that characterises the Aussie–Pom relationship, a fair bit of banter went on and, over time, our friendship became increasingly competitive.

This rivalry found an outlet in two ways: on the cricket pitch, where the bowling could become every bit as spirited (though not, sadly, as accurate) as on the 1933 Bodyline Tour; and in the way we flew our aircraft off the island.

Both of us usually stayed at a place called Le Chaland, a beautiful beachside hotel/resort around a kilometre from the end of the airport's main runway.

It was here, and at a number of restaurants around the island, that we raised our glasses, noisily often, to the many things that divided our great nations, listening to our two captains as they sought to outdo each other with tales of some of the characters they'd flown with during the war.

Because of our proximity to the airfield, it wasn't unusual for departing aircraft to fly over the hotel.

Due to the needling nature of our rivalry, these take-offs, over time, got lower and lower.

For us, in our VC10, we didn't really see this as an issue, because we had a daytime departure slot and – not to put too fine a point on it – we could see where we were going. The Aussies, on the other hand, always took off at night – at around 9.30.

On this particular night, we broke with tradition and ate in at the hotel. We made the mistake, however, of allowing the Qantas boys to know this.

As we were settling into a round of after-dinner drinks, we picked up the whine of the departing Qantas jet.

The 707 was a noisy beast at the best of times, but as the whine of those JT3s grew and grew, the plates and glasses started to rattle – and then the room began to shake.

As the Qantas jet streaked over, I thought it had taken the roof off.

The next morning, our captain got a telex from his Qantas oppo informing us that our little competition was over. As he'd taxied in and shut down, a ground engineer had walked onto the flight deck and said: 'Jeez, mate, do you know you've got half a tree stuck in your undercarriage?'

The top of the tree hadn't significantly damaged the aircraft, but even so . . . after that, the cricket matches continued, but the barnstorming stopped.

The afternoon we'd received the telex, as I lay on the beach in my hammock, I reflected on what had happened. A warm wind was blowing in off the ocean and I looked up.

I was a long way from Bournemouth, but through half-closed eyes I imagined my seven-year-old self gazing at silver specs flying eastward.

Even in the few short years I'd been flying with BOAC, I

knew – as we crews all knew – where our limits were: where you could do certain things, where you couldn't, as well as when you should and when you shouldn't.

Generally, the things you couldn't or shouldn't do were built on those 'foundations of understanding', as I'd come to think of them by now. As crew, you knew why a rule governing a particular do or don't was there.

Take speed, for example.

If you flew an aircraft beyond its maximum design speed – what we referred to as V_{ne} (never-exceed speed) – there was a good chance you'd damage the aircraft or worse. Nor should you make an approach in weather conditions that were below your minima because, as proved fatally the case with that Canadian Pacific DC-8 at Tokyo-Haneda (not that it was their fault – they'd been directed to land by ATC), you were jeopardising the aircraft and everyone on board.

There were limits, in other words.

Within this rules-based system, there was, however, no strong logic sometimes for why you didn't do certain things, other than that it wasn't good professional practice.

I peered into the blue a little harder and imagined I saw the silver underbelly of a jet high in the sky on that fateful day over Mount Fuji.

There were no rules to say that you could or couldn't use an airliner as a sightseeing platform. As a professional, however, you knew – or ought to know – that it was bad practice.

Low flying fell into the same category. Flying over a hotel at a thousand feet was no less dangerous than flying over it at a hundred, *provided you didn't hit the hotel* . . .

Bad practices crept in, and were exacerbated over time, unless you stuck rigidly to the standards of your professional code. How, I wanted to know, had our code warped and buckled in the way I'd just seen – I'd just been party to?

As I ruminated on this, a thought came. The pilots and navigators of the new age – an age epitomised by the VC10, 707 and DC-8 – were still dominated by the wartime generation. Part of me had loved being scared by Geoff Morell as he had pointed our VC10 at the Cliffs of Moher – but Morell had been a bomber pilot. Bernard Dobson, the captain of the 707 that had crashed on Mount Fuji, had also been a bomber pilot – a Dambuster, even, just as Chopper Knights had been. These people weren't in and of themselves *unsafe* – they were brilliant flyers. But they were the product of a different age. A different culture. A culture in which you'd had to go above and beyond what was expected of you to get the job done – whether that meant flying against impossible odds to sink a heavily armed battleship or destroying a dam.

We who'd had no experience of wartime flying had been reared via a more rigid process. As the ex-bomber pilots had started to retire, the airlines had needed new blood – the Hamble generation, if you like – a generation for whom the jets of the new age were in *our* blood; we'd known nothing else.

And yet, a hardy core of those wartime pilots still occupied left-hand seats.

That was where I needed to get myself to, I said to myself.

On my return to the UK, I went for a very early navigator's course, got on one, passed it and became a P2 – getting my second-in-command early; before I'd even become a senior first officer. It was a route that suited me – but it didn't suit everyone.

Becoming a relatively young, junior P2 opened other doors. I now had an opportunity to become a co-pilot instructor on the VC10, something I really wanted. The opportunity finally presented itself in 1976, after I'd been flying the aircraft for seven years.

But that was the year Concorde came into service with British Airways, the airline formed two years earlier from the merger of BOAC and BEA.

The writing had been on the wall for BOAC for a number of years. In the late 1960s, a body called the Edwards Committee had recommended the formation of a 'British Airways Board' to bring BOAC and BEA under one roof.

The rationalisation was originally intended to maintain the identities of the two airlines as distinct government-owned, public-sector enterprises. However, in the interests of best value for money for the travelling public, safety, and the equitable distribution of traffic rights and operating resources, the two were brought together formally as the fully merged 'BA' on 31st March 1974.

I needed to decide whether to put my name forward as a VC10 co-pilot instructor, something that would have been very rewarding and have opened up a number of career paths; or bet everything on Concorde.

I couldn't do both – but, equally, I might bid and get neither; the rules said that I had to make a decision and opt for one. On one level, it was a tough choice. On the head level, in terms of what, logically, it was better for me to do, I should have bid for the instructor's course because I stood a reasonably high chance of getting it. And it would have meant I'd be financially better off – you were paid more as an instructor than as a Concorde junior pilot. And it was a qualification that would have served me well for the rest of my career – once an instructor, always an instructor.

But, on the heart level, it was a no-brainer. I was on that

beach in Bournemouth looking up. I was in that study room at Hamble, listening to Raymond Baxter's awed tones on the TV as Concorde had touched down at Fairford.

I put my name down for the Concorde course at the first available opportunity.

Part of me had reasoned that this aircraft – far more than the VC10, the 707 and the DC-8 – was the airliner that would set the threshold for the next age of flight; an age in which a new generation of aircrew would sit on the flight deck. An era of rules-based flying that would, by necessity, leave the seat-of-the-pants types behind.

9

Concorde had entered service with BA and Air France at the beginning of that year – January 1976. The Concorde story up to this point had been long and chequered.

In the mid-1950s, various research and design studies into supersonic flight at the UK's Royal Aircraft Establishment at Farnborough had, by 1956, led to the formation of a Supersonic Transport Advisory Committee.

In the days when Britain still had a sizeable and largely independent aviation industry, the STAC comprised nine British airframe manufacturers, four engine companies, leading research establishments, several government departments and BOAC and BEA.

The STAC's first meeting, on 5th November 1956, resulted in hundreds of written submissions covering the feasibility of a supersonic transport aircraft.

These dealt with multiple aspects of the challenge ahead, including materials, systems, engines and the economic and social ramifications, not the least of which was the impact sonic booms would have on populated areas – this at a time when it was envisaged that such an aircraft would fly supersonically over land.

Meantime, supersonic transport aircraft studies were under way in France, the USA and the Soviet Union.

The French initially were set on developing a supersonic airliner based on their highly successful Sud Aviation Caravelle;

this for use on European and African routes – a medium-range design France remained wedded to for some time.

In the USA, Boeing, Douglas and Lockheed were all involved in SST, or supersonic transport, studies. Intense competition from the UK, France and the USSR had resulted in the unprecedented decision by Washington to develop an SST utilising government subsidy channelled through the National Advisory Committee for Aeronautics – NACA, forerunner of NASA – with Boeing eventually selected to design and build an SST at the end of 1966, the Boeing 2707.

Britain had been courting both the USA and France as its supersonic partner. However, with constant prevarication by the US, coupled with what turned out to be, by the early 1960s, a marked similarity between the British and French designs, the British government eventually opted to partner with France. The decision was tied into gaining Paris's support for Britain's entry into the 'Common Market' – the European Economic Community, forerunner of the European Union.

In November 1962, the Anglo-French agreement marking the beginning of Concorde's development was signed. By now, the greater part of the British airframe industry had merged into one nationalised company – the British Aircraft Corporation.

BAC, it was decided, would design and build the fuselage from the nose gear forward, including the aircraft's distinctive droop-snoot and visor; and from the trailing edge of the wing rearward, including the fin and rudder.

Aérospatiale, the prime French contractor, would take the fuselage centre-section and, crucially, the distinctive delta wing; this being the main area of aerodynamic breakthrough that would see the aircraft transition across the critical Mach 0.85 to Mach 2 'bridge': an engineering feat that would be vital in taking it to its design cruise speed of twice the speed of sound.

The other vital factor in its performance was, of course, the engine.

Bristol Siddeley Engines and French engine-maker SNECMA had already agreed by the aircraft's launch that they would cooperate. The Bristol Olympus engine, which had been designed for the RAF's Vulcan V-bomber, was viewed as an engine of considerable potential. A more powerful version of the Olympus was already in development for the RAF's new strike fighter the BAC TSR-2 (which would later end up cancelled). Why re-invent the wheel?

With development, most critically via the addition of a 'reheat', or 'afterburner', a section of the engine at the back that injected fuel into the engine exhaust stream, expanding the exhaust gases and boosting thrust by as much as half again, it was envisaged that the Olympus would have the necessary power and performance to meet the highly challenging design specification. Reheat was used at critical junctures of the aircraft's flight profile, most notably on take-off, to help lift the heavily loaded aircraft into the air, and to punch the aircraft through the transonic regime to Mach 1 – and on towards its cruise speed of Mach 2

Under the bilateral agreement, it was decided, therefore, that Bristol, which would end up purchased by Rolls-Royce in 1966, would be responsible for the basic engine and SNECMA for the 'bit at the back': the jet pipe, reheat section, nozzles and thrust reverser.

The redesigned powerplant emerged, ultimately, as the Rolls-Royce/SNECMA Olympus 593.

Britain, it was agreed at the same time, would have 60 per cent of the engine work; France, 60 per cent of the airframe.

Because of its responsibility for the engines, it was decided that Britain should also have design and development authority for the engine intakes and control systems.

Under the terms of the treaty, however, both countries would have an equal share of the production work.

It then became a matter of finding a name.

In 1963, French President Charles de Gaulle referred to the aircraft as 'Concorde'. The British, throughout the development and prototype construction phase, stuck dogmatically to the anglicised version – 'Concord' – until, in 1967, Tony Benn, Minister of Technology for a Labour government that had once threatened to cancel the aircraft, announced that the British would revert to the French *Concorde*. In the resulting nationalist uproar, Benn claimed that the suffix stood for Excellence, England, Europe and Entente (as in 'Cordiale'). When it was pointed out that key parts of the plane were also being developed and built in Scotland, Benn – in a neat piece of bilingual wordsmithing – rejoined that it also stood for Écosse, the French word for Scotland. He would later, tongue in cheek, say it might also have stood for Extravagance and (cost) Escalation as well.

The flight of the first prototype – 001, assembled by Aérospatiale in Toulouse – took place on 2nd March 1969. The second prototype, 002, assembled by BAC at Filton, Bristol – the flight I had watched while studying for my final exams at Hamble – took place a few weeks later. In October of that year, on its 45th test flight, 001 broke the sound barrier for the first time. In November the following year, both 001 and 002 attained Mach 2, twice the speed of sound.

In July 1972, BOAC ordered five Concordes and Air France four. By now, the US SST programme had been abandoned. The Boeing 2707, which would have been almost three times the size of Concorde and capable of almost Mach 3, had existed as little more than an impressive full-scale wooden mock-up in a Boeing hangar in Seattle at the time of the cancellation order – and even that had by now been broken up. But if Concorde's manufacturers thought they were about to be presented with an uncontested market for supersonic transport aircraft, they were in for a shock. In January 1974, as economic conditions globally worsened – triggered by a fourfold increase in the price of oil following the announcement of the OPEC embargo the year before – Pan Am cancelled its order for Concorde. Other cancellations followed.

Worse, in a sense, was to come.

In December 1975, the US House of Representatives voted by 199 votes to 198 to impose a six-month ban on Concorde landing in the USA on noise pollution grounds – though everyone knew this was a fudge. With no SST of their own, America was pissed, to put it mildly, that, for the first time in a generation, it had been outwitted technologically. Transatlantic destinations – New York and Washington D.C. in particular – had always been envisaged as the primary routes for the aircraft – the jewel in any airline's crown.

With Concorde set to enter service the following year, a monumental lobbying campaign ensued by the French and British governments, the two aircraft manufacturers, Air France and the fledgling British Airways, to overturn the decision.

With the ban still in effect, the two airlines had to find other routes for their inaugural Concorde services – BA opting for Bahrain and Air France for Rio de Janeiro, Brazil, via Dakar in West Africa. On 21st January 1976, those two services began – the historic moment recorded in a simultaneous take-off – timed to the second – from London and Paris. BA hoped that Bahrain would later form the first leg of a journey that would take Concorde on to Singapore and Australia.

The next month, the US government overturned its ban, giving approval to BA and Air France to commence a one-a-day service to Washington and two-a-day to New York. Washington's Dulles Airport was owned and operated by the Federal Aviation Authority, so no problem there. JFK Airport, on the other hand, was owned by the Port Authority of New York, which refused landing rights, despite the Federal government's ruling.

Operations to Washington-Dulles began simultaneously by BA and Air France on 24th May 1976, with departure times adjusted so that the two aircraft landed at the same time. They taxied ceremoniously around the airport ending up nose-to-nose. They then dipped and raised their noses and visors in salute. And the press loved it.

The ban in and out of JFK was immediately challenged by the two airlines, the Port Authority eventually conceding it would make its judgement after results of noise studies at Heathrow, Dulles and Paris Charles de Gaulle.

The people chosen for each Concorde course were determined – and, to a degree, made the determination themselves – according to three things.

First, seniority. A Concorde 'course' had emerged with slots for five captains and five co-pilots, presenting me with an opportunity to bid for a co-pilot slot. Lots of people were able to bid; not everyone, however, chose to.

One of the conditions of bidding as a Concorde co-pilot was that you had to agree to a minimum of five years because of the cost of training. That meant you couldn't become a captain elsewhere – the five-year commitment was sacrosanct. Going for Concorde at that time, therefore, could seriously disrupt your chances of a command – this being the second consideration.

The third was that you had to be 'technically suitable', a nebulous criterion that meant someone, somewhere had some input as to whether they thought you were up to the job. The problem, however, was that the process was opaque – the technical suitability requirement was never spelled out.

The decision was further complicated by the way Concorde pilots were remunerated. Bizarrely, in those days, they earned less than pilots of other aircraft.

While basic salary was determined by the type of aircraft you flew – the categories being propeller-driven planes like the Viscount, jets like the VC10 and the 707, and 'big fast jets' like the 747 and Concorde – salary wasn't the whole story.

A chunk of your take-home pay came from your allowances on the route. If you went away on a 747 trip, you might be gone for twelve days, meaning you could earn a lot in allowances. Whereas the longest trip away on Concorde was two days – and the differential could be up to £10,000 a year – this when £10,000 was worth a great deal more than it is today. So, a lot of pilots who could have bid didn't for that reason. A lot of others didn't because they wanted to see the world – and the irony, at that point, was that Concorde was restricted to a limited number of routes.

But the reason people *did* bid was, in essence, for the same reason I chose to: they loved flying – and the opportunity to fly Concorde trumped everything else. I put my bid in during that long, hot summer of 1976.

And immediately hit a problem.

Because of the differential in pay, Concorde's instructors decided they did indeed have a raw deal. They were instructors on Concorde, the new toy, a complex aircraft to instruct on, and they were taking home far less than their counterparts on other aircraft.

So, there was a bust-up between the instructor group and the company – effectively (although it was a little more nuanced than this), the instructors went on strike.

All courses were cancelled for the next eight months – well into 1977.

When they resolved the issue (with what amounted to a fudged pay deal), the courses started up again. It might have been reasonable to expect BA to have gone back to all the people who'd previously registered, but it didn't; it went to a rebid – everyone who'd bid before had to bid all over again.

This insanity went one step further in that my group was now told that we couldn't rebid because in the time it had taken to sort out the pay dispute, we'd become too senior – we were now 'too close' to getting a command.

This, frankly, was rubbish. There was no way, whatever the rules stated, we'd get a command within five years, and we pointed this out. Anyone who'd spent any time reading the runes knew it would take far longer.

In the end, accepting this and all our other points, we were allowed to rebid, and almost all of us who'd been accepted for the original course were green lit for training. The financial side for me and my colleagues was neither here nor there. We were talking the difference between a lot of money and even more money. We did it because we wanted to fly the only supersonic passenger aircraft in the world.

Correction: the only supersonic passenger aircraft in the *Western* world.

After America's tryst with the SST, culminating in the cancellation of the Boeing 2707 in 1971, the Soviet Union had succeeded in developing its version of an SST. The Tupolev Tu-144 had been dubbed 'Concordski' by the press because of a strong resemblance to its Anglo-French counterpart.

The Tu-144 had made its maiden flight on 31st December 1968 and had entered service with Aeroflot seven years later, in December 1975 – operations being restricted to a route between Moscow and Almaty, Kazakhstan, three time zones to the east.

Any hopes of selling the Tu-144 abroad had been dashed largely for the same reason Concorde's sales had been so drastically affected – the sharp rise in world fuel prices following the OPEC oil embargo.

But the Tu-144 had also crashed tragically and spectacularly in full view of the world at the 1973 Paris Air Show at Le Bourget. The demo crew, it transpired, had been ordered to outperform Concorde. The Tu-144 departed from its normal flight regime and, as the crew fought desperately to bring it back under control, under the stresses of the attempted recovery the left wing broke off, prior to the disintegration of much of the rest of the airframe. The pieces fell on to a suburb of Le Bourget killing all six on board and eight on the ground.

Concorde was an altogether different kettle of fish. Here was an aircraft that had pushed almost every single boundary when it came to aerospace innovation. She was a source of national pride, a symbol, too, of international cooperation, and we had every faith that she was going to be a winner.

But it was clear, too, that learning to fly this aeroplane would be no piece of cake.

Of the three groups of aircraft – turboprops, jets and big jets – captains on those aircraft with an aspiration to fly Concorde came, by definition, from Group 2, the category that comprised the VC10 and 707. The 747 hadn't been around very long, so if they were on the 747, they couldn't have jumped to Concorde even if they'd wanted to.

So, they all came from Group 2. If you stepped up from Group 2 to Group 3 you were in for a basic pay-rise, excepting that you'd not get as many allowances on Concorde as you would on the 747.

Concorde being the new toy, to begin with it was the very senior guys who bid. But being senior, they hadn't done a conversion course for ten or more years and, in a lot of cases, had forgotten something fundamental: *how to learn.*

It doesn't take long to learn how to learn again, but the Concorde course was so intensive that if you got a week behind – taking a week to learn how to learn again – then you'd be struggling from then on. The failure rate amongst this group of people was 50 per cent. Which, in itself, wasn't good.

But the sting in the tail was that if you were on a Group 2 aircraft and you bid for a Group 3 aircraft – Concorde or the Jumbo – and you failed the course, you went back to your Group 2 job, back on to a lower salary, and were then stuck there for three years.

So, it became obvious fairly soon that a lot of people weren't prepared to take that risk. They *were* prepared to sacrifice money to fly this wonderful aeroplane, but they didn't want to get locked in for three years if they failed.

As a result, all of a sudden, the bidding profile went from the most senior captains to the most junior, because the junior captains had nothing to lose. And it would be three years or more before they had the seniority to bid on the 747 anyway. One guy I knew bid on Concorde having been a captain for only eighteen months.

From then on, a perception that the course was very difficult prevailed. The course *was* difficult in one sense, in that the things you had to know – not so much the complexity but the sheer volume of *stuff* – was immense: on Concorde there were far more systems than on any other aircraft, you were training for a whole new flight regime and the aircraft had radically different handling qualities to all other civilian aeroplanes.

A conventional course to retrain a pilot from one aircraft type to another took two months. On Concorde, it took six.

Four of those months – the two-month technical course and the two-month simulator course – were residential in Bristol at the British Aerospace Concorde production facility, Filton.

You had to have a very understanding family to put up with your being away for all that time – coming back on a Friday night knackered and disappearing back down to Bristol on Sunday afternoon – and, probably, during much of the two days you *were* back home revising – cramming for tests – because at the end of every week there was an exam.

In short, you had to be very dedicated to want to sacrifice ten grand a year and be away from your friends and family for all that time.

The advantage to all of this was that the Concorde fleet got a group of pilots who were technically very competent and capable – otherwise they wouldn't have exposed themselves to the risk of staying stuck where they were for three years. Their commitment to the aeroplane was second to none.

The downside of this was that it was the root cause of the people on the Concorde fleet getting a name for themselves as a bunch of elitists.

This, in my experience, in no way reflected the reality of the situation – but it was a perception that stuck.

And, later, at a time when the aircraft needed all the help it could get, it was one that would resurface as a knife that would be plunged and twisted into the aircraft's bright, white underbelly.

But all this was many years into the future – and would come as one of the unforeseen aftershocks of the Paris crash.

For now, I was as happy as I could be. I was twenty-seven years old and set to train on a supersonic aeroplane – a phenomenal piece of world-beating engineering – that had already been hailed, and rightly so, as an icon.

10

The first two months of the course – the technical component – were spent in the classroom.

To begin with, we had to get used to being trained not by BA, but by the manufacturer – British Aerospace – at Filton, in Bristol, which is also where the Concorde simulator was housed.

As well as teaching you how to fly the aircraft, the sim was employed on the technical course to help you understand its systems. It was Monday to Friday, 08:30 to 17:30, 'chalk and talk', with an exam before we went home every weekend. We were tested on the subjects we'd covered during the week, as well as everything we'd covered thus far as the course went on – so the cumulative effect was considerable.

At the end of it all, we were heading for a final exam that had 390 questions, covering every aspect of the aircraft. The exam wasn't even multiple-choice – we had to provide detailed answers to every question and the pass-mark was 80 per cent – only, naturally, we all strived to do better.

Providing you passed the technical course – after a week at Heathrow for briefings on safety equipment and procedures – it was on to the next step: the simulator course. This comprised a further two months at Filton and nineteen four-hour simulator 'details'.

For this, we were 'crewed up' – we were part of a captain, co-pilot and flight engineer team operating as what was known as an 'established crew'. This was far from normal in the airline business – and not what I was used to at all on sim training either. Normally, on a 'conventional' simulator course, you would probably switch pilots and navigators two or three times. In the military, flight crew are geared to flying as a permanent tight-knit team – it engenders positive morale and is vital when you're flying combat operations, when you need to be able to read the mind of your fellow crew members. In the airline business, you fly with whomsoever you're rostered to fly with on any given trip.

The benefit of working as an 'established crew' was that you could all help each other as you melded together as a team – a form of collaboration that was actively encouraged and had been especially important at Hamble. You knew from the syllabus what was coming up and to get ahead you'd mug up. You'd then get together as a team and mug up some more, because a captain's perspective on a particular issue would often be quite different from a co-pilot's or a flight engineer's; and all three of you would likely benefit from the discussion that brought the various perspectives out.

In carrying out an emergency drill for a particular failure, for example, an engineer might pick up an important facet of the issue we might not have seen; or the converse: the engineer might not have appreciated the event's significance from a 'tactical' point of view – how it played out when you're sitting behind the controls.

This was a really good aspect of the Concorde course – something that the other members of my crew and I had not experienced before. The other two were a captain, Dave Brister, and a flight engineer called Tony Brown.

The work bonds you form as a normal flight crew are very different, of course, from the bonds you form in, say, an office. This was what made the established crew deal on the

Concorde sim course unusual. I was the junior member, but the standard operating procedures encouraged you to get rid of the seniority/juniority gradient in the way that you all worked together.

There was a very good reason for this.

The 1972 crash of the BEA Trident near to the King George VI Reservoir at Staines had happened, in part, the enquiry had established, because of the way the captain and co-pilots had interacted with each other on the flight deck. As I recalled all too well from my deep-stall training at Shannon, the Trident – BEA Flight 548 *Papa India* – had gone into a deep-stall soon after take-off.

This should have been entirely avoidable thanks to the provision of the stick-shaker: the device installed on all T-tail aircraft after the 1-11 crash in Wiltshire; on which I'd familiarised myself thanks to Chopper Knights' tuition in Ireland.

In the investigation that followed the Trident crash, it was established that other factors – human factors – had contributed significantly in addition to the technical issues behind the stall.

The captain was Captain Stanley Key, an ex-RAF Second World War combat pilot. An hour and a half before *Papa India* took off, destination Brussels with a full complement of 112 passengers, Captain Key was involved in an extremely heated row with a first officer named Flavell in the crew room at Heathrow.

Its trigger had been a strike, ongoing at the time, by twenty-two 'supervisory first officers' (SFOs) – co-pilots – who, like the striking Concorde instructors four years later (albeit for different reasons), were aggrieved at their low pay and high workload relative to other pilots. In other airlines and on other aircraft, the SFO role was usually undertaken by flight engineers. In BEA, SFOs were P3s. On the Trident they sat on the third flight-deck seat, operating the systems and helping the captain, a P1, while tutoring the trainee co-pilot, a P2.

The SFO/P3s, denied the chance to fly the aircraft, were hit by a loss of pay, which had led to their going on strike.

The argument between Captain Key and Flavell was over another threatened strike – this one in concert with other airlines over the wave of recent hijackings that had been occurring – and, according to an eyewitness, it had bordered on the violent. Whilst Captain Key apologised to Flavell, it was speculated he might have carried some of his anger on to the flight deck of *Papa India*. Key was fifty-one and had 15,000 flying hours, 4,000 of them on Tridents. His P2 was twenty-two and had only started flying routes six weeks earlier. The P3/SFO was twenty-four and had over 1,400 hours, 750 of them on Tridents.

Another pilot, Captain John Collins, an experienced former Trident first officer, had been allocated the observer's seat under a practice known as 'deadheading' – he was part of a crew tasked with bringing a BEA cargo-carrying aircraft at Brussels back to Heathrow. Collins' two other crewmen were in the cabin, the extra weight adding 24 kg to the aircraft's permitted maximum gross take-off weight of 41,730 kg. This was comfortably offset, however, by the fuel that the aircraft had burnt off while taxying.

All of these factors, and the details underpinning them, would become critical to the events that followed.

Key took off into squally conditions – low cloud and driving rain at approximately 16:09. The aircraft left the ground at an indicated airspeed of 145 knots, quickly reached its safe climb speed of 152 knots – V_2 – and the gear was retracted.

At 16:10, at an altitude of approximately 700 feet, the Trident entered cloud and Key throttled back under standard noise abatement protocols.

Seconds later, he retracted the flaps from their take-off setting of 20 degrees.

After passing 1,500 feet, *Papa India* was cleared to 6,000 feet *en route* to Dover via the non-directional radio beacon, NDB, at Epsom, one of the UK's standard navigational aids, and thence to Brussels. During the turn to the NDB, the airspeed dropped to 157 knots, 20 knots below its target speed.

At 16:10:24, one minute and 54 seconds after the aircraft had begun its take-off roll, at almost 1,800 feet, the leading-edge slats were retracted – 63 knots below the safe retraction speed of 225 knots – prompting visual and audible warnings of a stall on the flight deck, followed a second later by a stick-shake and, a second after that, by a stick-push, which deactivated the autopilot. Slats are aerodynamic surfaces at the front of the wing, which, when lowered into the slipstream, allow an aircraft to fly at higher angles of attack and therefore to fly slower. On *Papa India*, this action triggered a warning horn that persisted for the rest of the time that the aircraft was in the air.

Approximately two minutes after take-off, and after several more stick-shakes and stick-pushes, the Trident went into a deep-stall, with its nose pitched up 31 degrees and its airspeed below the minimum speed that could be accurately recorded on the aircraft's airspeed indicator – 54 knots. This was identified subsequently from the aircraft's 'black box' – its flight data recorder.

In a deep-stall, its tail blanketed by the vortices coming off the wings, there was nothing short of a miracle that anyone on the flight deck could have done – although the subsequent report indicated that the crew had tried desperately to save the aircraft. *Papa India* was falling out of the sky by now at 4,500 feet per minute in an almost flat attitude. Its engines still roaring for a velocity it could not now attain, it pancaked, nose-up, at a speed that would have been sedate even for a car.

The aircraft broke up on impact. There was no fire. The only survivor found at the scene subsequently died.

The accident remains the worst in UK aviation history, outside of the terrorist-induced crash of Pan Am Flight 103 at Lockerbie in 1988.

In the enquiry that followed, an urgent call for cockpit voice recorders ensured that CVRs soon became mandatory on all commercial aircraft flight decks.

Whilst it was never determined who had selected the premature retraction of the leading-edge slats, a focus soon fell on the unorthodox mix of crew on the flight deck as a key contributor to the crash – and the fact, as had been determined by a pathologist, that Captain Key had suffered some kind of 'cardiac event' between his argument in the crew room and the moment of impact.

It was also suggested that Captain Collins's presence in the observer's seat might have distracted *Papa India*'s regular crew, although there was evidence, too, to suggest that it had been Collins who had been attempting to redeploy the slats at the moment of impact. A fireman – the first person to enter the shattered flight deck – had found his body slumped over the centre console and his earphones in the right-hand footwell, indicative of the fact he'd been attempting to intervene during *Papa India*'s fall from the sky.

We will, however, never know.

The fact that it led to what was known as 'cockpit resource management', later called 'crew resource management', was one of the crucial outcomes of the crash. CRM is a structured decision-making process in which everyone on the flight deck understands what they and everyone else is supposed to be doing through SOPs: standard operating procedures.

Papa India's legacy was active measures to get around the 'seniority gradient'.

Under CRM protocols, senior captains and co-pilots were trained thereafter to be challenged by junior crew on the flight deck – as long as there was no personal element to the challenge. In 1977, five years after the *Papa India* crash, CRM was not yet endemic across the industry, but it was – due to the complexity of the aircraft and its systems – something that had been introduced onto Concorde via the established crew training we underwent at Filton.

This would later prove to be another crucial piece of the jigsaw in our attempts to understand the events that led up to the crash of Air France 4590 at Gonesse.

The distinguishing feature of the two Concorde simulators (one British, one French) was that they were built around genuine Concorde nose sections – they were the twenty-first and twenty-second airframes to roll off the production line, only there was nothing aft of the flight deck bulkhead.

The UK version took the nose and integrated it with a six-axis motion platform and its associated systems – the visual suite, computer and interface – built as a joint venture between Singer Link-Miles and Redifon Flight Simulation. The image generated by the camera was displayed on TV displays in front of the cockpit windows.

Air France took the other nose section and integrated it with its own computer systems and visuals package. This had been developed and built by French simulation company Le Matériel Téléphonique. The Air France simulator was based at Toulouse.

The Concorde simulator we trained on at Bristol was a step-change in sophistication compared to the VC10 simulator I'd 'flown' at Cranebank.

Although still based on the same crude (by today's standards) visuals system – comprising a Redifon closed-circuit TV camera moving over three large model landscapes mounted vertically on the wall in the adjacent room – the six-axis motion system was an order of magnitude improvement over the VC10's one-axis system. The jacks allowed for motion in the pitch, roll, yaw, heave, sway and surge domains. The surge – or 'lunge' function – even managed to provide some semblance of 'g'. Although it was nothing like the g-force sensation you'd get in a real aircraft, the overall effect was pretty realistic.

Of the nineteen four-hour details we got on the sim, it was the seventeenth that was critical: the detail that would test our overall competence as a crew.

Even if you passed a detail as an individual, you could still fail if you didn't gel as a crew.

This was the principle behind the established crew system. Concorde was the only aircraft where the person at the systems panel had to be a fully qualified engineer. BA had flight engineers on their other long-haul aircraft, but not on the short-haul ones. And that was a 'BA choice'. Many airlines flew the same types, like the earlier 747s, with a pilot at the panel. Not Concorde. My pilot's qualification actually said that it was not valid unless there was a fully qualified engineer looking after things – and me!

The benefit of having a flight engineer train with you is, first, they're not after your job and, second, they've got a vested interest in ensuring that no 'bastard pilot' is going to kill them – thus, they are quite prepared to intervene should they feel the need to. Experience told us, even after the *Papa India* BEA Trident crash, that a junior flight engineer was usually more inclined to ask a senior captain 'what the heck he was doing' than a very senior co-pilot.

Flight engineers had a very different perspective – their point of view was engineering-focused and so was highly complementary to the pilot's.

The downside, if you were a pilot, was that during the Concorde conversion course, we had to know the aircraft's systems. And if you happened to be a co-pilot, you had to know them almost as well as the flight engineer. If the engineer was taken ill, it was down to the co-pilot to clamber out of his seat and take over at the systems panel.

And Mach 2 could be an unforgiving place.

On most airlines – and on the short- and medium-haul side of BA such as the Trident and the new Lockheed TriStar – pilots had routine experience of sitting at the systems panel and thus their knowledge of that particular aircraft's systems was constantly being refreshed. This wasn't the case with Concorde because of the uniqueness of the flight engineer's status on that aircraft.

At Filton, therefore, I had to make a real effort to understand the aircraft's systems panel – as a co-pilot, I'd be the one who would have to operate it if anything happened to the engineer. There was a lot to learn.

Here to teach me was a man named Norman Todd, another ex-bomber pilot. He had flown with the RAF in India on four-engine Liberators, attacking Japanese overland communications links, including the infamous Burma Railway. After his forty-second and final mission, in which he'd been forced to make an emergency landing when one of his engines failed, he moved to a training role. He left the RAF in 1946 and joined BOAC, flying on Atlantic and African routes, and was amongst the first of the airline's pilots to convert to the 747.

In 1973, Norman was appointed to assist in the Concorde flight test programme and had piloted the first commercial Concorde flight from Heathrow to Bahrain in January 1976. He had also piloted BA's first transatlantic Concorde flight to Washington and, in November 1977, he would fly the Queen and Prince Philip home from Barbados at the end of their Silver Jubilee tour of Canada and the West Indies. This was a record-breaking flight of 4,200 miles covered in three hours and forty-two minutes at an average speed of 1,134 mph.

Norman wasn't out to fail me – he wasn't out to fail anyone. No one is trying to trick you or catch you out. It was in everyone's interest that you passed the simulator course, but you had to put the work in – and then some.

By the end, I knew how to handle the aircraft and perform in every emergency situation we might ever be expected to encounter. But however good the simulator was, it was no substitute for the real thing. The time soon came when I needed to fine-tune my handling skills by flying Concorde for real.

RAF Brize Norton, Oxfordshire, was the Royal Air Force's prime strategic air transport base, located around 75 miles to the west of London, close to the city of Oxford. With its long runway, relative isolation from large population centres, and – for all that – its handy proximity still to Heathrow, Brize had been designated the ideal airfield for Concorde flight training. One of the key parts of the syllabus was a requirement for the trainee conversion pilot to learn how to perform 'touch and gos' – landing and taking off again seconds after the wheels have greased the tarmac. Anyone who has witnessed fighter aircraft doing this knows that it is an extremely noisy procedure. Touch and gos on Concorde with its four afterburning Olympus 593 engines – each generating in excess of 38,000 pounds of thrust with reheat – took this to an altogether different level. With apologies to the inhabitants of Brize Norton, Carterton and Witney – the villages and towns bordering the base – it had been decided, in concert with the RAF, that this was the best place for us to train.

I woke up on that summer's morning – Monday 8th August, 1977 – at a hotel in the neighbouring countryside filled with excitement at what lay ahead. The aircraft we were to train on had been taken out of normal passenger service and deployed to Brize a few days earlier with a full complement of support staff – a ground engineering team and the spares and maintenance tools needed for the aircraft's two-week deployment. I had driven down from my home near Heathrow the evening before.

After a full breakfast, I settled down with another co-pilot who would also be trained that day in Concorde basic handling. Two pilots and two flight engineers would deploy per flight, each of us conducting around half a dozen 'circuits and bumps' before we swapped seats. Our flight was scheduled for 4:30 that afternoon. Another crew would go through exactly the same exercise before us.

It was like Shannon all over again, only Shannon seemed like a different era.

My instructor at Brize was Captain Keith Myers, a nice guy, not a Chopper. As we headed out of the hotel to the minibus that would take us to the base, I felt the butterflies that had accompanied me at other big moments like this. This time, though, they felt different.

After the obligatory security checks at the base gate, we were waved through and instructed to head for the 'terminal' – Brize was, in some respects, little different from an airport; the RAF's major hub for arrivals and departures to and from points all over the world. As we pulled up in front of the building, I heard the familiar roar of Rolls-Royce Conways – and moments later a VC10 in the colours of RAF Air Support Command roared overhead.

We were led through into one of the two briefing rooms that had been set aside for us. Through the window a shaft of sunlight broke through the clouds, catching something dazzlingly white on the ramp in the distance. I squinted against the brightness and saw Concorde. *Weird*, a voice in my head said, *unlike other aircraft, you never attached a definite article to the name*. It wasn't *a* Concorde or even *the* Concorde out there, but just *Concorde*. The voice then said: *And, today, they're going to let you fly that beautiful beast.*

I looked at my watch. The next four hours couldn't go fast enough.

The first crew left the building and were driven out to where the aircraft was waiting.

I heard the engines start up; watched as they taxied Concorde away until she was no longer in view. Then, there was that roar – the roar of those Olympuses. Seconds later, she came into view again as she thundered past us down the runway, pulling up into the sunshine. Then she was gone again – and I was left with the ringing of the four afterburning engines in my ears. Ten minutes later, I heard the first hint of a howl – a noise you picked up long before you heard other jets. Then I could see her, established on final approach with her nose fully drooped, landing gear down and her bright lights on. She looked like a giant white pterodactyl, swooping towards the ground.

Then an almighty roar as the Olympuses opened up again. A 'go-around'. She soared away, raising her nose and gear and was quickly gone.

Then I remembered: the first approach of the first detail is always a 'go-around', so the pilot can practise what are called 'baulked landings'; to get the beast down to her maximum landing weight for the next approach and landing.

About six minutes passed before she reappeared. That same vision – of a prehistoric bird coming home to roost – then she shot past us on the runway, the throttles opened up and she pulled up into the sky again – the first of several touch and gos that the trainee crew before us would be practising.

Finally, I heard the reassuringly familiar sound of the Olympuses in reverse-thrust, signalling, at long last, that their time was up. Another interminable hour went by while the ground crew turned her around. Then it was our turn.

Keith Myers came with us on the bus out to the aircraft. 'How are you feeling?' he asked, as he puffed on a cigarette in the seat next to me. I stared straight ahead. The red and blue Union Flag flash on Concorde's fin grew until it filled most of the windscreen.

'Excited,' I replied. And then, worried that that hadn't sounded too professional, I added: 'I mean I'm looking forward to it, obviously, but . . .'

Keith interrupted me. 'How old are you, Mike?'

'Twenty-eight,' I said.

'Well, then, you *should* be excited.'

'Were *you*?' I asked. These instructors were still like gods to me. They were so cool and calm. Excitement wasn't something I felt them capable of.

'Of course. I'm here to teach you how to fly her, but I want you to remember this day, too. This is going to be a memorable moment in your aviation career. Take the time to soak it all up. Take the time to *enjoy* it.'

The bus pulled up a short distance from the airstairs that led up to the door behind the flight deck.

I clambered out and gazed up. I hadn't fully appreciated just how big she was till now. And tall. I had seen and heard Concordes around Heathrow, of course, but I had never been up close to one. The aircraft was almost as long as a 747; her tall, spindly legs holding her high off the ground. She made the RAF VC10s and C-130s on the ramp around her look archaic by comparison.

I followed Keith up the airstairs. He turned left for the

flight deck. I turned right. I wanted to see the cabin before I settled into the right-hand seat.

BA's Super VC10s could accommodate 170 to 180 passengers in a high-density seating configuration, with six seats abreast. Concorde was 100 seats, two abreast, but due to her narrower, pencil-like fuselage, the overall effect, compounded by the small size of the windows, was of a much more confined interior than I'd anticipated – even having been warned about it.

I sat down in seat 1A, having been told this was where royalty and VVIPs sat. Very comfortable, and oodles of leg room. Then, I jumped up, about turned, and headed back for the flight deck.

Even though I'd acclimatised to Concorde's cockpit from my simulator sessions, entering the real thing was quite an experience.

Every inch of space from the floor up, forward of the engineer's position, was taken up with switches and dials. And there was a smell – the smell of hot electronics – that made the real thing very different from the sim.

I eased myself into the right-hand seat and went through my checks – all of them from memory – while Keith and the flight engineer went through theirs.

When we were done, in a break with tradition, Keith asked if I'd like to head back down the airstairs to join the engineer for his walkaround inspection.

We started at the nose.

Aside from that delta planform, Concorde's nose was her most distinguished and distinguishing feature. Unlike other airliners, which used leading-edge slats and trailing-edge flaps to lower their safe handling speeds at take-off and on approach, Concorde's overriding requirement to cruise supersonically required her to be as streamlined as possible. This had led to her breakthrough design feature – her signature delta wing – known as an 'ogival delta' due to the shapely ogival or S-shaped curve of the leading-edge.

But delta wings brought other challenges.

The way the wing generated lift at slow speeds was via a phenomenon known as 'vortex lift'.

All swept wings create vortices – swirls of air at their wingtips. Concorde's wing, however, created much larger, slower-moving vortices during high-angle-of-attack flight – most critically during landing – and these, typically, formed over the wing's entire upper surface. This created an area of low pressure above the wing – a suction force a bit like a Dyson cleaner – that generated the lift required to keep her in the air at slow speeds.

At normal attitudes, when she was in fast, straight and level flight, a delta generated lift like any other wing.

But those high angles of attack, her designers realised, plus the fact there was around 25 feet of nose ahead of the cockpit, would make it impossible for the pilots to get a clear view of the runway on landing. Various ideas were toyed with to solve the vision issue – including a periscope – until someone came up with the idea of a 'droop-snoot': a nose that would lower by as much as 12.5 degrees on landing. The take-off angle was set at a far less radical 5 degrees.

The nose also included a protective visor for the windscreen that was deployed during all other portions of the aircraft's flight profile. The visor was made of special heat-resistant glass one-and-a-half inches thick that could cope with the high temperatures and aerodynamic stresses associated with sustained supersonic flight. The nose was raised and lowered hydraulically via a four-position locking lever to the left of the first officer's control column.

I strolled a few feet further aft until I stood before one of the great unsung heroes of Concorde's innovation story: the aircraft's air-intake system.

Whilst the development of Concorde's reheated Olympus 593 engine was an achievement in itself, a considerable ancillary effort had been made to allow the engine to operate across the spectrum of speeds it would fly at – from take-off to Mach 1, thence to a cruise speed of Mach 2, and then back down to subsonic speeds for landing.

This cycle Concorde had to do repeatedly and reliably every day of the week for what was expected to be an operational life of thirty years-plus.

The Olympus was a turbojet engine, as opposed to a more modern turbo*fan* – the difference being the engine's 'bypass ratio': the air that went around or bypassed the 'core' – the guts of the engine: diffuser, combustors, nozzle and the various stages of the turbine that drove the compressor – versus air going *through* the core.

The Olympus was chosen not just because it was already a mature, ten-year-old design (with, crucially, a mature engine *core*, the hardest part of an engine to develop) when the Concorde project was launched, but because its narrow width and high exhaust speed made it the most efficient engine for the portion of the flight envelope where Concorde would spend most of her time: Mach 2.

Even today, the Olympus 593 Mk 610 variants installed on production Concordes remain the most efficient engines in the world at Mach 2 and above. At slower speeds, however, the 593 610s were inefficient, which is why we needed to get Concorde to her design cruise speed as rapidly as possible.

Dynamically, though, her problems were only just beginning there, because no conventional turbine can accept air into its core at twice the speed of sound.

The task facing Concorde's designers, therefore, was to radically slow the air entering the compressor stage at the front of the engine – from Mach 2 to a more manageable Mach 0.5 (or from about 1,350 mph to less than 500 mph).

The solution was what I now took a careful look at: the engine's ultra-long, computer-controlled air intakes.

The intakes were governed by a highly advanced digital control system for regulating the amount of airflow entering the engine – a system that was, as far as the crew was concerned, entirely hands-off. Air entering the engine – the right amount of air at the right moment of the aircraft's flight profile – was regulated automatically and was critical to Concorde's cruise mode at an altitude of 60,000 feet. Remarkably, to sustain Mach 2 at this height, she didn't need her afterburners – the punch-power of the engines that we usually referred to as 'reheat' – due to the high efficiency of the Olympus at such speeds and the highly streamlined design of Concorde's airframe.

The air intake was a big, boxy, rectangular structure that ran a full 11 feet from the intake's lip to the engine's initial compressor stage.

A pair of moving doors – ramps – were installed in the roof of the intake and these were lowered into the airstream to regulate the intake flow. Two smaller doors served to let in more air or to spill it when not needed.

All in all, the integration of the intake system with the engine was a miracle of design that enabled Concorde to perform across all modes of her flight envelope; something, it was rumoured, the Americans and Russians had really struggled with.

At take-off, the ramps were positioned wide open to allow as much air as possible to flow into the engine. All jet engines need to gulp air at the beginning of their take-off runs, and the Olympus was no exception. Reheat was used from the start of the take-off roll for roughly a minute, reheat increasing the thrust to each engine by around 5,000 pounds – from 33,000 pounds without reheat (known as 'dry power') to 38,000 pounds with.

Due to the noise this created – much more than a turbofan and/or a turbojet in dry power – we needed to keep the use of reheat to a minimum.

Reheat radically helped, however, to get a fully laden Concorde off the ground. It also helped to overcome the very large increase in drag associated with the so-called 'transonic region' – the part of the envelope where the aircraft broke through the sound barrier.

At Mach 1.3, the ramps lowered to form a series of shock waves that slowed the air on its journey from the lip of the intake to the first stage of the engine's compressor. The computers – remarkable in their own right for the day – sensed the changes in air pressure and temperature properties to the wave patterns to adjust the position of the ramps.

Were Concorde to suffer a catastrophic engine failure at Mach 2, the non-rotating turbines would present themselves as a near-flat surface to the direction of the airflow, presenting a potentially catastrophic amount of drag.

To prevent this from happening, the ramps – again under the control of the computers – would drop instantly, diverting some air over the top of the engine.

A 'spill door' would also fall open, dumping air out of the intake's underside.

Due to this unique arrangement, Concorde behaved unlike any other aircraft that suffers an engine-out at high speed.

All aircraft yaw into a dead engine because the engine goes from being a producer of thrust to a producer of drag – the aerodynamic force that opposes an aircraft's motion through the air.

As I'd practised on the simulator and for real on the VC10, this caused the wing on the opposite side of the failed engine to travel faster and produce more lift, causing it to rise.

At subsonic speeds, Concorde would be affected in the same way; at supersonic speeds, with the ramps clamped shut and air spilling out beneath the affected engine, the effect was the opposite: it was the wing with the dead engine that rose as the spill-door air acted as a jet pushing the wing up, causing Concorde to roll away from the dead engine.

The effect was reduced by auto-stabilisation but still needed pilot input to counter.

Not for the first time, I marvelled at the ingenuity – the sheer breadth of all the systems knowledge and its implementation that came together in this aircraft.

Next, her undercarriage. I'd found taxying on the VC10 challenging enough, but Concorde's nosewheel was over 60 feet behind the point of her nose, and hence about 35 feet behind the pilot, making turning on the ground unique to her. There were two wheels on the nosewheel bogie; the tyres were made by Dunlop for BA variants and by Goodyear for Air France's.

Tyre pressure was set at 190 PSI (for a comparison, think of the 30-odd pounds per square inch we averagely pump into the tyres of our cars). The strength of the tyre lay in the cords built into the rubber beneath the tread.

Built up layer upon layer, the cords, made of nylon, formed a casing strong enough to sustain the aircraft's landing load every time it thumped on to the tarmac. The cord casing also had to be strong enough to resist all that air pressure – the 190 PSI being the other thing that gave the tyre its strength.

Unlike the tread of a car tyre, aeroplane tyres weren't composed of criss-crossed treads – they had grooves in them; this to allow them to skid until their rotation speed matched the speed of the aircraft. This resistance is the reason aircraft tyres emit a puff of smoke at the moment of touch-down.

Tyres rarely, if ever, blew because they were *over*-inflated. They were another of those engineering miracles – as overlooked as Concorde's air-intake system – that people rarely gave thought to. Even so, they could fail – usually when they were under-inflated or overloaded. Casings could blow and treads could come off.

Twenty-three years later, it would be just such a failure that would lead to the loss of the aircraft.

The right main landing gear, which I walked to next, comprised a giant four-wheel bogie unit, whose principal design function was to absorb Concorde's landing energy every time she touched down. Because of the aircraft's high angle of attack on landing, 11 degrees as standard, the rear part of the engine nozzles, which protruded just aft of the wing trailing edge, were at risk of striking the runway. To get around this, Concorde had long, extendable main landing-gear struts. Were each main gear assembly to retract normally, the two sets of wheels would have struck each other. Thus, the landing gear retraction sequence necessitated a shortening of the struts followed by a lateral swing into the belly.

After giving the tyres an appreciative kick, I strolled aft to the aircraft's elevons – the control surfaces on the trailing edge of each wing.

There were three elevons per wing: inner, middle and outer. These controlled the aircraft in pitch and roll.

Because the aircraft lacked a tailplane, the elevons provided her with nose-up and nose-down control, like a conventional 'elevator', combined with roll control like a conventional 'aileron'. Morph the words and you had an 'elevon'.

The fin, meanwhile, which was divided into an upper and lower rudder, provided yaw control.

Beneath the fin was the aircraft's final distinguishing feature: her tail-bumper landing gear. This was a small retractable wheel that was designed to protect the tail during take-off and landing due to the aircraft's tail-dragging, low speed, high angle of attack. Were it not for the tail-bumper, under certain conditions – even with those long, tall, mainwheel struts – the engine reverse-thrust mechanism would scrape the tarmac on landing.

The walkaround took the engineer and me a little more than ten minutes. We paid particular attention to the static wicks – discharge devices for the build-up of electricity on the airframe during flight – and the pitot heads, antennas and sensors, looking for signs of wear or damage.

Subliminally almost, too, I took in the other things I'd been trained to do over the years: the condition of the ramp, the proximity of other aircraft, the speed and direction of the wind.

The inspection over, the feeling of excitement returned.

I walked slowly back to the nose, mounted the airstairs, turned left onto the flight deck, breathed in the smell of the cockpit avionics again and slipped into the right-hand seat, next to Keith, who was already busy with his pre-flight cockpit checks.

12

I set my seat and began my memory drills, going around all the instruments and systems I was supposed to check in order – radios, weather radar, flight controls – as Keith and the trainee engineer started to go through theirs. I set up the inertial navigation system, the INS, by inputting the aircraft's exact position on the ground, thereby allowing it to plot our location (this, again, being the pre-satnav era) throughout the time we'd be in the air.

It took me around five minutes to do my share of the twenty-five or so checks. I could see Keith out of the corner of my eye methodically going through things on his side of the cockpit too.

When we finished, he turned to me. This was the part that was routine to me – the part where, as a crew, we talked through exactly what each of us would be doing on take-off. This time, clearly though, it was a little different.

'After engine-start, we'll take a long taxy out to the end of the runway,' Keith said. 'Taxying this aircraft is different from taxying any other because the nosewheel is so far behind us. So, there are a few party tricks to learn.'

'Tricks?' I said.

'We'll do a few turns, including some right-angle turns, so you can get the hang of it. Happy with that?'

'Happy with that,' I confirmed.

'We'll do all our checks before we get to the runway so that the only thing we have to concentrate on is the take-off and our entry into the circuit pattern. We'll make sure we're not holding anyone up and, if necessary, sit at the end of the runway for as long as it takes – there's no hurry – while we talk things through with air traffic. We'll fly five or six circuits in the pattern – take-offs and landings – so that you can get the feel of the aeroplane. All clear?'

'Yes,' I acknowledged.

'When we've completed all the checks, what are you going to do?'

'When we're ready to go, I'll say "three-two-one . . . *now*." I'll open up the throttles to full power and then, as we go down the runway, I'm expecting you to make certain calls.' I asked him then what calls he was planning to make.

'I'll call "airspeed building" and, at decision speed, "V_1". What are you going to do at V_1?'

'I'm going to take my left hand off the throttles and place it on the control column, because that'll be the last chance we get to stop.'

Keith nodded. 'That's going to come up fast – we don't weigh much and we're going to be using full power for take-off. The acceleration is going to be different from anything you've experienced – and, as you know, it can't be replicated on the sim. The next thing I'll say is "rotate". What do you do then?'

'I'll pull back on the control column and pitch us to our pre-bugged value of eighteen-and-a-half degrees nose-up,' I told him, referring to our pre-computed attitude for climb-out at V_2 on three engines based on our relatively light weight. 'It'll take between five and six seconds for us to do that,' I added, meaning it would take up to six seconds of steady pull-back to get us to our 18.5 degrees pitch attitude. 'You'll then call "positive climb" and I'll call for "gear up". As we climb ahead through 500 feet, I'll call "reheat off" and the engineer will switch off the afterburners. At a thousand feet, I'll call "climb power", we'll throttle the engines back and select the auto-throttles to maintain 250 knots.'

I glanced at the auto-throttle window on the coaming ahead of me. 'Two-fifty knots is pre-selected.'

Keith nodded again. 'What I'll ask you to do then is climb straight ahead and level off. Then, when we've got everything settled, we'll turn right into the circuit pattern and brief as we go around the circuit. There'll be time in the circuit to brief on all the items I want you to do. All clear. Any questions?'

I shook my head.

'Good. Let's take five minutes. Then we'll go to engines-start.'

Starting the engines couldn't happen before the arrival of the load sheet.

On a scheduled flight, the load sheet was produced by a central computer. For our training flights, it was produced by the crew who'd fly the next detail – we returned the favour when it was their go. Producing a load sheet the old-fashioned way – using manuals – was a little like learning slide-rule navigation in today's era of satnav. But it was something we had to do.

The way Concorde was loaded was key to the way she handled and flew.

The load sheet gave the crew all the critical information they needed for any given flight: take-off weight; the amount of fuel loaded; and where the aircraft's centre of gravity would be with everything and everyone on board.

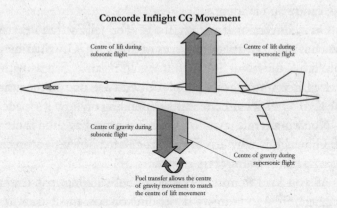

Concorde Inflight CG Movement

Centre of lift during subsonic flight

Centre of lift during supersonic flight

Centre of gravity during subsonic flight

Centre of gravity during supersonic flight

Fuel transfer allows the centre of gravity movement to match the centre of lift movement

Now, as I studied the load sheet on the Brize ramp, it accounted for the aircraft's dead weight on the tarmac, the weight of the next crew and where they were sitting in the cabin, where all the catering was stored (even though it was just a few sandwiches and flasks of tea and coffee on this flight), the aircraft's planned weight at take-off and her all-important centre of gravity position.

All-important, in fact, was an understatement.

You didn't need to be an aerospace engineer to see that Concorde was a thoroughbred. But her thoroughbred lines came at a cost. And that cost was the system that ensured that her centre of gravity – her C of G – was always in the correct position, no matter what portion of the flight she found herself in.

Like her Olympus engines and her computer-controlled air-intake system, both of which catered to Concorde's wide speed range, an elaborate system was designed and developed to ensure that her C of G was always in the exact right place at the exact right time – those critical moments being take-off, cruise and landing.

On a conventional aircraft, the C of G falls within a certain band. If the aircraft were a seesaw, the C of G would be its fulcrum – the point at which it would balance. But as with a seesaw, if you've got two people one end and one on the other, the C of G of an aircraft shifts depending on how it's loaded.

No two aircraft, even of the same type, have the same C of G, which is why it is always calculated down to a precise point when the aircraft is empty.

As you load items on board – fuel, catering, passengers, baggage, crew and cargo – depending on how much each item weighs and where it's loaded, the C of G will move forwards or backwards. This calculation is so sensitive it can vary according to the weight and location of individual passengers.

On a conventional aircraft you're OK to go provided the C of G falls within an 'allowable band'. This band can be quite long. This is because you, as the pilot, use a device called a 'trim-tab' to adjust the aircraft's aerodynamic balance within the band.

Trim-tabs are small secondary flight controls attached to the trailing edge of primary flight control surfaces – elevators, ailerons and rudders.

Using a wheel, or a button on the control column, a pilot sets the position of the tab prior to take-off in a way to ensure that the force he or she applies to the stick at the moment he or she lifts the aircraft off the runway – that moment of rotation, or V_R – is as near uniform as possible for each flight.

Throughout the flight, as the passengers, cabin crew and trolleys move around the cabin, and you burn off fuel, you maintain the C of G using trim-tabs. On most aircraft, this is done automatically when the autopilot is on.

On Concorde, though, it wasn't like this at all.

Trimming Concorde's C of G using trim-tabs would have been out of the question because it would have moved the affected control surface into the slipstream where it would cause drag – a force contrary to the direction of thrust that slows the aircraft down.

On a conventional aircraft, drag matters because it affects fuel-efficiency – but it isn't all that critical. A small percentage increase in overall drag is inconsequential relative to the available thrust and the amount of fuel a large widebody like a 747 can carry.

Not so with Concorde. On Concorde, streamlining for supersonic flight was the be-all and end-all of her existence. Streamlining was everything.

Efficiency on Concorde was heavily predicated on the relationship between the C of G and the aircraft's centre of lift. The 'C of L' is the point where the sum total of all lift generated by an aircraft – via its wings, control surfaces and even the fuselage – acts as an aggregate force on the airframe. As the aircraft moves faster, the C of L shifts aft. The distance between the C of G and the C of L becomes a 'lever-arm' akin to the lever-arm on that ol' seesaw.

Without the luxury of those drag-inducing trim-tabs, Concorde needed to find another method to optimise the length of that lever-arm.

Too small a gap and the aircraft would become unstable; too big and it would become unmanoeuvrable.

The method her designers came up with in the end for maintaining this balance was brilliantly ingenious: they did it by moving her fuel in flight.

Fast-forward now twenty-three years to the crash.

By the end of that terrible first day, we'd acquired certain crucial pieces of information.

We knew the aircraft had run over a piece of metal that had been dropped onto the runway by a Continental Airlines DC-10 that had departed moments earlier. The narrative then unfolded as follows: the front inboard tyre on the Air France Concorde's left undercarriage bogie had exploded on hitting the metal, sending blocks of rubber and pieces of cord casing, fired by the force of air pressurised at 232 pounds per square inch, into the underside of the wing. The tank had then ruptured, the fuel that then streamed from the tank had caught fire, and the rest constituted a tragic milestone in aviation history: the moment, effectively, that the dream of supersonic air travel died.

But this, we found out in the weeks *after* the crash, was very far from the real story, which was woven around several key elements – one of the most important being the aircraft's centre of gravity and that ingenious fuel-transfer system: the means by which Concorde shifted her C of G in flight.

Concorde contained a series of fuel tanks distributed throughout the airframe. Numbered 1 to 11, this network consisted of main tanks, engine feed-tanks and trim-tanks.

As Concorde went faster, to adjust the C of G rearward in flight, fuel was pumped from the tanks at the front of the aircraft to the tank at the back: to Tank 11 beneath the fin.

Around 10 tonnes needed to be transferred in this way to keep her trimmed.

The fuel was always loaded in a particular way as well. The fuel hoses were plugged in and a number tapped into the computer that told the system how much fuel you needed to put on board. The bowser operators did something similar, so that the two numbers tallied. As the fuel was being

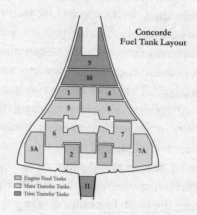

Concorde
Fuel Tank Layout

Engine Feed Tanks
Main Transfer Tanks
Trim Transfer Tanks

pumped, the aircraft distributed it to the right tank at the right moment by opening and closing a series of valves and pumps that kept the C of G in the same place – this being determined by the kind of trip you were to make.

On a transatlantic flight, we were talking a lot of fuel: up to around 95 tonnes (today, by contrast, for my training flight out of Brize, we were loading a meagre 35 tonnes). The computer assigned quantities of fuel to each tank, guiding the entire operation while it was happening – this was not just to manage the C of G, but, until the final 4 tonnes went aboard, to keep it constant and to avoid upending the aircraft on the ramp: not enough forward and too much at the back and the aircraft would end up on its tail.

Today, at Brize, we were operating at Concorde's 'light-weight' C of G setting.

The aircraft had three C of G settings: light, medium and heavy.

Critical to the aircraft's fuel-transfer system in flight – how much fuel to pump from front to back – was an exact knowledge of where the C of G was before take-off.

Concorde's designers decided that the best way to determine this was to do away with the 'acceptable band' that was standard on conventional aircraft and go for a fixed C of G point for take-off. This also drove the way Concorde was loaded before moving off-ramp. Clearly, there had to be some flexibility in the loading process, because no two loading procedures were ever the same.

But prior to the moment Concorde began her take-off run, that little bit of flexibility had to be driven out of the system. Fixing the C of G point on take-off was sacrosanct.

On a lightweight profile – such as today's out of Brize – that point was 53 per cent: 53 per cent of the distance from the aircraft's nose to its tail.

At lightweight, by definition, you didn't have so many passengers on board or so much luggage – you could put them roughly where you wanted.

As you got heavier, however, the seats got occupied and the hold filled, meaning your capacity for controlling C of G via loading alone was markedly reduced.

This set a medium weight C of G point at 53.5 per cent.

But then, not long after Concorde entered service, it was decided that for really heavy-weight settings – i.e. for transatlantic flights with a full load of a hundred passengers and their baggage – a new C of G setting was needed: 54 per cent.

But here was the catch.

When refuelling to a full load of up to around 95 tonnes, as you hit the 89-tonne mark, all the tanks were full – except for one: Tank 11 under the fin.

Tank 11 had about 4 tonnes in it by now, but still had

capacity. Any more fuel you put on board at this point, therefore, *had* to go into this tank.

But because it was at the back, it would shift the C of G aft.

Which meant another little trick was needed to shift it forward – to where it had to be for take-off – by pumping fuel forward from Tank 11 as you burnt fuel off during the taxy.

The four engines got their fuel directly from four main feeder-tanks.

Each engine had its own: Tanks 1, 2, 3 and 4. You couldn't pump fuel forward until you'd burnt enough fuel out of those four feeder-tanks to make space for fuel from Tank 11. The pumps could pump fuel quickly – they were powerful pumps. Getting it forward was limited by the rate at which the engines drank fuel from the feeder-tanks – this under a process called 'pre-take-off burn-off'.

Eventually, you got to a point of diminishing returns – where every kilo of fuel you put in Tank 11 had to be burnt off pre-take-off in order to get the C of G to where it needed to be. This, of course, made absolutely no sense.

There was, however, another option – and that was to fool the system.

This being my first flight, I was unaware then that such a thing was possible.

Fooling the system, though, would become central all those years later in our journey to understand what had gone so tragically wrong at Gonesse.

13

As soon as she began to roll, I wondered if I'd swapped our plane for a sports car. The Olympus 593s were so powerful that, even at engines idle – at our Brize-focused light weight – Concorde leapt out of the blocks and kept accelerating. You took your foot off the brakes and, *bloody hell, she went.*

I quickly learnt that applying the brakes was an issue: too much and they overheated, and then you couldn't take off. Braking was vital in the event of an emergency-stop at the V_1 decision-point.

So, before I got anywhere near the runway, the first thing I had to learn was how to brake when taxying at light weight. You did this, Keith reminded me, by letting her go faster than you'd normally be comfortable with. A normal aircraft you probably taxied at 10 to 12 mph, but with Concorde you let her get up to 25 mph before you dipped the brakes. Then you went through the cycle again, letting the speed build before applying the brakes again. You could even alternate right and left brakes to give each an extra bit of time to cool.

Steering was even more counterintuitive. That long, high nosewheel was about 38 feet behind you – and the two sets of mainwheels even further: 95 feet. Keith showed me how to do a couple of right-angle turns before we got to the runway threshold.

When we'd been through all our before-take-off checks and cleared with air traffic, Keith turned to me.

'Right, Mike. Using full power and reheat, I want you to take off and climb straight ahead.'

This wasn't quite as straightforward as it sounded. Concorde being a noisy beast, it had long been determined the best way to minimise noise was to use full power and reheat for all take-offs to allow her to get up and away quickly. If you lived next to the airport you'd be none too pleased, but in terms of the overall noise footprint, lots of power early was the preferred procedure.

So far, so good, but all this was for a normal take-off on a normal day. Full power was designed for a fully laden Concorde at her maximum take-off weight of 185 tonnes – *and* on the assumption that after V_1 only three of the four engines might be working.

For this, my first take-off, I was going to have all four Olympuses turning and burning with 60 fewer tonnes to worry about – or hold us back.

'Are you ready?' Keith said.

'Ready,' I replied, tersely.

He nodded. 'OK, then take off, climb straight ahead and level off at 2,500 feet.' He then glanced at me, smiled mischievously and added: 'If you can.'

I held his gaze for a fraction of a second, as I pondered this weird qualification. *If I can? What did he mean: If I can? I'd been flying for years.*

'Three . . . two . . . one . . . *now!*' I banged the throttles. No matter how many times I'd done this in the sim, banging the throttles still felt wrong. You banged them on a 747 and those big turbines would probably explode. But on Concorde, the computers were so advanced they could handle almost anything you threw at them. All four engines wound up instantly at the same predetermined rate.

The force that pushed me back in my seat was as unexpected as it was phenomenal.

In the simulator, you got nudged back in your seat thanks to 'pushback' – the 'shove mode' that was supposed to get you used to the sensation of 'g'.

But this was altogether different. The g-force going through my body, though perhaps only twice normal, was like nothing I'd experienced before. And it *increased* as we went down the runway.

For a moment, my senses were flooded. I refocused as Keith called our speed: 'one hundred knots, power-check, V_1, rotate . . .'

I pulled back on the stick – 18.5 degrees pitch-up, I suddenly realised, was a long way. On a conventional aircraft, you'd pitch up less than 10 degrees. As I pulled back, I felt the g's come on again – this time in a way the sim couldn't begin to mimic.

And that power! The 18.5-degree climb-out was predicated on our losing one engine. With four engines, we were going up like a rocket. I eased back on the stick some more, knowing that when we reached 500 feet those reheats were going to get switched off, necessitating a nudge to the nose to compensate for the loss of power. But we weren't there yet – so I pulled the nose up to 22 degrees, then a fraction more.

Our climb speed was limited to 250 knots because the maximum speed for retracting the undercarriage was 270 and we needed some buffer to keep us on the right side of safe.

That thought had no sooner left my head than the reheats came off, pitching me forward.

I pushed the nose down to maintain our speed, reminding myself that in another 500 feet we had to reduce power again and this would require another 'de-rotation' – dropping the nose yet again – to maintain our all-important climb-out speed.

All this had happened in less than a minute-and-a-half since brakes-off and it wasn't over.

To level off at 2,500 feet, you had to pitch the nose forward towards 5 degrees as you passed 1,800 feet. I was aware that I had been doing little else except push-and-pull since we'd lifted off the runway – and hoped I'd not been too ham-fisted about it. The last thing you wanted was to 'bunt' – pull negative 'g' – because on a fare-paying flight you'd be scraping the passengers off the roof.

The trick was to do all that to-ing and fro-ing on the stick as gently as possible, I'd learnt in the sim, to leave all on board with the impression it was one long, seamless climb. The auto-throttles had clicked in as the reheats had winked out, so at least I didn't have to worry about that side of things. Any moment, I told myself, and we'd be up to 2,500 feet, our level-out altitude.

I checked the altimeter – and couldn't believe my eyes. We were at 4,000!

I glanced at Keith, who was laughing his head off. 'Happens every time.'

'It does?'

'Happened to me,' he said. 'It's a wild ride, isn't it?'

It certainly was. I felt like I'd just ridden an unbroken colt in a rodeo.

The circuit is a preconceived pattern in the sky – a racetrack around the airfield that would be mine to practise on for the next hour. I coaxed us back down to 2,500 feet, realising as I did so that ATC must have anticipated this, because the airspace above had temporarily been cleared of all other traffic.

The purpose of the initial exercise was to line us up for an approach and landing.

Still at 250 knots, and by now several miles clear of the airfield, I initiated 30 degrees of bank until I had got myself on to a reciprocal with the Brize runway, now visible out of my right-side window. As I came abeam the upwind end of it, I reduced speed to 210 knots and started to descend.

Forty-five seconds later, I turned in at right angles to the runway and reduced airspeed to 190 knots

At 1,200 feet, I turned on to final approach.

From this moment, our final approach speed – 'V Reference' – was set, allowing me to fly the glideslope and pinpoint our touch-down spot on the runway, which was looming larger by the second through the windshield.

Thanks to the droop-snoot, despite our high angle of attack, the forward visibility was good.

I focused on the PAPI lights to the left of the runway. The Precision Approach Path Indicator was a bank of four horizontal lights that told you if you were above or below glideslope. Too high and all the lights showed white. Too low and they showed red. If you're 'on glideslope', you get two whites and two reds. Nice and simple.

Except PAPIs were set up for a conventional aeroplane.

Because Concorde's undercarriage was set way back from the cockpit – and the PAPI was set up for approaches by conventional aircraft – this required some adjustment. Instead of the point directed to me by the red and white lights, I needed to land a little deeper, which meant opting for a pink-white colour indication instead.

Keith was doing what all good training captains

do – leaving me to my own devices. He was a calm presence who'd intervene only if he needed to.

Meantime, I was racking my brains from all my sessions in the sim.

The sim's motion dynamics – how it responded to your input on the controls – were excellent. But the visuals package – the TV image projected in front of the windshield by the camera moving across the model landscape – in no way simulated what was actually happening outside the windows. That level of sophistication wouldn't arrive on simulators for several more years.

Then, just before Keith said, 'Go around,' I remembered that, on this first approach, we were going to practise a baulked landing. That would involve applying full power, climbing, accelerating, raising the gear and the nose part way and climbing back up to 2,500 feet. Having been startled by the aircraft's 'heaven-bound' performance on take-off, to say I was apprehensive was an understatement. Keith gave the 'Go around' command and, this time, we didn't use reheat to climb out.

But she still went like a homesick angel.

I contained the beast to just 2,800 feet before drifting back to 2,500 and going around the pattern again.

Back on to final approach and into 'the slot' – on the runway centre line and onto the glideslope at the right speed. Here, now, if anything, because I was getting all the right visual cues from the outside world, I was finding it a little easier than I'd anticipated. But I knew I had to be wary – I had to avoid at all costs the very tempting idea that landing Concorde was like landing any other large plane.

For a start, we were coming in fast – 162 knots versus the 120-knot approach I'd have made on a VC10. Over a third faster. I was still flying the same three-degree descent path, but because of Concorde's speed, the rate of descent was a lot quicker – 800 feet per minute instead of 600 feet. *So*, I told myself, *you're landing deeper than normal, but not* that *much deeper . . .*

Because of our rate of descent, the speed at which the houses and trees were whipping by beneath us would have been semi-hypnotic if I'd allowed myself to be distracted by them. My visual scan-path was a constant series of eye movements between the outside world and my instruments – especially my airspeed indicator and the ILS indicator, two needles on a display in front of me that, in addition to the cues I was getting from the PAPI, were guiding me to my touchdown. The ILS – a device at the ends of the runway that beamed localiser and glideslope data to sensors on the aircraft – was essential for days when airports were 'socked in' with fog, but not today. Today, I had all the data I needed from the outside world to make my landing – almost too much, in fact.

The angled windows at the side of the cockpit, because we were 13 degrees nose-up, gave an impression the aircraft was in level flight. Keith reminded me the window was canted, giving a deceptive impression of our attitude.

He gently reminded me now, too, about the nosewheel.

You had to 'land' the nosewheel, flying the sim had told me.

In a conventional aircraft, even when you were descending at a comparatively leisurely 600 feet per minute, you had to markedly arrest your rate of descent before you touched down. There's a little bit of an air cushion that develops beneath an aircraft as it gets close to the ground called 'ground-effect'. This will arrest your rate of descent, but not by much. You'd still hit the ground at a rate of 400 feet a minute – more than enough to jar a few bones.

To avoid this, therefore, you 'round-out' – reduce the rate of descent by pulling back on the control column, adjusting the elevators to lift the nose of the aircraft, just prior to touchdown. This process meant your rate of descent when your wheels hit the tarmac was a nice and gentle 10 feet per minute.

And because you've closed the throttles, the nose will drop from its 5 to 6 degrees nose-up attitude as well, enabling it pretty much to land itself.

Concorde didn't perform this way at all.

Approaching the airfield at 162 knots, descending at an elevator-like rate of descent of 800 feet per minute, nose pitched right up, the aircraft's delta wing generated a huge cushion of air close to the ground – a massive amount of ground-effect. So much so that, 50 feet from the ground, I could actually hear it as a rush of air – a sound I'd not heard on any other plane. On a VC10, ground-effect would reduce your rate of descent maybe by around 15 to 20 per cent.

On Concorde, it was 90 to 95 per cent.

In effect, this meant the aircraft didn't want to land.

Keith had pre-briefed me on this with a fascinating fact. With a little proficiency, you could actually fly the aircraft straight into ground-effect, he'd told me, and it would pretty much stop descending all by itself.

I wouldn't be risking that today.

Today, I needed to pull back on the stick for a different reason. As well as generating ground-effect, Concorde's aerodynamics generated a push-down force on the nose.

To stop it thumping into the tarmac, you had to exert a gentle pull-back on the stick; and I needed to manage that back-pressure even more carefully as I reduced the throttles, because closing the engines increased the effect on that nose-down moment. In effect, I was maintaining my pitch-attitude to stop the nose dropping – not because I needed to flare the aircraft on touchdown.

All of this was totally counterintuitive – the opposite, almost, of everything I'd learnt.

Now, as the main wheels touched down, with the tail-bumper wheel just off the ground, I focused on landing the nosewheel. It was already dropping, but I didn't want to land it heavily, so I reduced the amount of pull-back.

Talk about patting my head and rubbing my stomach . . .

More weird cues now from the side windows, which, because the nose was dropping, were accelerating an impression – coming to me from my peripheral vision – that the nose was about to bury itself into the runway.

'Looking good, Mike . . .' Keith's reassuring voice told me to trust my instincts; that everything was OK.

With a reassuringly firm, but still gentle, thump, the nose-wheel greased the tarmac.

Instinctively, my hands moved to apply reverse thrust, my feet ready to hit the brakes. But, fortunately, before I committed, before I got a physical rap over my knuckles from Keith, I remembered that this was a touch and go.

'Go,' Keith instructed me.

Instead of closing them, I slammed open the throttles to full power – full dry power. We were still barrelling along the runway at 120 knots, enough energy for us not to need to go to reheat. I watched the speed build past 165 knots, 180, 185 . . .

'. . . rotate.'

At 190 knots we were airborne again, to repeat the process a few more times before the detail was over. Fortunately, as time went on, my technique improved, and the landings got smoother, until it started to become second nature.

I'd just completed 'Detail 1'. Over the next several days, I went through many more – poor weather details, crosswind details, details in which the landing aids were switched off, simulated engine-out details and various combinations of potential failures.

By the end of my time at Brize, Keith sent a report off to the Civil Aviation Authority and a week or two later, I got a piece of paper that said I was now a fully qualified Concorde pilot. There was a rider, however. The licence, it said, was only valid if I had a fully qualified, licensed Concorde engineer sitting on the flight-deck behind me.

There was another catch, too. Although in the eyes of the CAA I was fully qualified to fly the aeroplane, in the eyes of my employer I wasn't yet.

As far as BA was concerned, I wouldn't be fully qualified until I'd flown through several more hoops.

There was a massive difference between flying Concorde at Brize and flying a route structure with passengers – and in the regime for which the aircraft had been designed and built in the first place: flying supersonically.

After Brize, I could look forward to all of this under the next segment of my training.

14

Boom!

We perceive a sonic boom as a sound, but it isn't; it's an overpressure.

Air can travel at a maximum speed – a speed that we have come to know as the speed of sound. The speed of sound is fixed – although its absolute value varies depending on temperature and altitude – because there's a limit to how fast air molecules can get out of the way of an object. On a standard day, at sea level static conditions, the speed of sound is around 760 mph or 1,100 feet per second.

The reason the speed of sound is referred to as a 'barrier' is because – in a supersonic aircraft – air molecules are being compressed and they can't get out of each other's way fast enough. In the end, in effect, they form a wall.

The air molecules pushed from the path of a Boeing 747 get out of each other's way with comparative ease because the aircraft is moving relatively slowly. But when a supersonic aircraft begins to punch through the wall, something odd happens – the molecules form into a cone-shaped bow-wave.

Concorde's overpressure was around 2.5 pounds per square inch, but this, too, depended on the temperature of the air, and the speed and altitude at which the aircraft was flying.

There's another misconception about sonic booms – that they only occur when an aircraft goes through the sound barrier. In fact, the boom is sustained – it's happening all the time the aircraft is going faster than Mach 1.

The cone that reaches a ground observer is heard as a boom – that is what those 2.5 pounds of pressure feel like to an eardrum. Their intensity will depend on distance, because overpressure dissipates – just as the wake of a boat does.

The further you get from the source, the more it weakens. Concorde's, however, was considered sufficiently intense for it not to be allowed to fly over land, apart from over very, very sparsely populated areas.

There were two reasons for this.

The first was its capacity to crack the odd window or break a piece of china.

The second was more nebulous but, in its way, more damaging. This was the damage potential – this time to the airline's reputation – of one of its aircraft rocketing over areas that were, perhaps, more socially challenged than others – where folks couldn't, let's say, afford to eat caviar and drink champagne on ice while they shaved a few hours off a journey from A to B.

Boom-boom!

A second cone came off the aircraft's tail as the air decompressed.

This was so close behind the first, however, that it was very often missed – but it was there, a force of nature; undeniable and unavoidable.

In November 1977, shortly after I got my Concorde co-pilot's certificate, we were granted permission from the New York Port Authority to start services to JFK.

To avoid the impact of the boom as we flew west, we needed to accelerate over the sea.

For the sake of fuel efficiency, we needed to do that as

soon as we possibly could, which meant spooling up – beginning our acceleration to supersonic speed – over the Bristol Channel.

In theory, Concorde could have trundled around at speeds up to Mach 1.2 with little disturbance below because between Mach 1 and Mach 1.3 a boom pretty much dissipates before it reaches the ground. But it was easier for regulators to issue a blanket ban when it came to supersonic flight over land than to set a complicated set of conditions. But there was another reason why this would not have worked – Concorde simply wasn't operating at maximum efficiency when flying below her design cruise speed of Mach 2.

As Concorde approached the speed of sound, the drag on the airframe rapidly increased.

It was for this reason we needed the afterburners – an additional 25 per cent power punch – to take the aircraft as rapidly as possible through this high-drag region – from Mach 0.95 at 28,000 feet to Mach 1.7 – towards her operating altitude.

Here, the genius of the Olympus engine was such that it was able to continue accelerating the aircraft from Mach 1.7 up to Mach 2 and 60,000 feet without the need for reheat.

Switching the reheats off saved a massive amount of fuel. One of the reasons Concorde was successful where the Russians' Tu-144 wasn't, was precisely because of this facet of her design. The Tu-144, we were told, had to have its reheats on pretty much all the time in order to sustain supersonic speed. This not only made it extremely thirsty, but noisy – so noisy, it was said, two passengers sitting next to each other couldn't hold a conversation when the reheats were burning. Whereas on Concorde, the reheats were noticeable, but not significantly so; and once they were switched off, the cabin wasn't appreciably noisier than the cabin of a conventional aircraft of its time.

As a trainee Concorde co-pilot, I knew all of this because I'd been through it on the sim. I'd been told, too, by some of the old hands, that going through the sound barrier was a bit of an anticlimax because there was nothing to feel.

No 'control reversal' like in the movies, no rattling, no shaking, no drinks spilt – just a smooth transition.

But I wanted to feel it for myself – and today, for the first time, I would.

I looked at my watch. In twenty minutes, I was due in 'Two Two One' – Briefing Room 221 in Terminal 3 – for a rundown from one of the best. Norman Todd, the veteran Concorde pilot who had presided over my simulator training at Filton and had commanded the very first BA Concorde scheduled flight in January 1976, was to be the captain on my first operational flight: the BA flight that was departing that morning, in just a few hours, from Terminal 3 to Washington Dulles. *Oh, the fun and excitement that lay ahead.*

Briefing Room 221 ran almost the entire length of Terminal 3's third floor. It was where the flight planners and the roster people sat – the latter being the people who made sure crews checked in OK and who chased them if they didn't. Cabin crews also gathered there. On this late August day, the sun streamed through the windows. Room 221 was always a hive of activity, filled with a background hum of purposeful chatter, and today was no different.

I walked over to Norman, who made the introductions to our flight engineer for the trip, Graham Tullier, and another pilot, John Massie, whose presence was required because I was still under instruction.

'First order of the day is for you to enjoy this, Mike,' Norman said with his customary largesse, as I took my seat for the briefing. 'This is the first time you're going to actually go supersonic in this aeroplane – the first time you're going to do what she was designed to do, and I want you to savour the experience, especially when we go through the sound barrier.'

He then added: 'I know that you'll have been briefed that it's sensation-free – and in a sense it is. But part of the experience will be appreciating just how sensation-free it is. And you'll be able to do that throughout the flight from the right-hand seat without too many distractions, because I'll be doing the take-offs and landings.'

I looked at him. I'd learnt at Brize how to fly the aircraft and, yes, I still needed to be taught about route operations and how to operate the aircraft supersonically across the Atlantic, but I had kind of assumed I would be doing the take-offs and landings today, so this came as a bit of a blow.

Norman must have sensed my disappointment, because he held my gaze and said: 'I repeat, Mike, this particular flight is more about enjoying yourself and having fun than ticking any particular boxes of the syllabus.'

As I thought about what he said, my feeling of frustration ebbed. This made sense. The purpose of my training from here on in was to learn how to fly this sophisticated supersonic aircraft efficiently, effectively, safely, on time, and economically. In other words, it was as much about understanding how to *operate* it as it was about handling it physically. I'd had briefings on how to fly transatlantic routes – I'd even flown a few transatlantic legs on the sim – but this would be a chance for me to acclimatise to the operating environment for real; a chance, too, under Norman's expert tuition, to test my own 'capacity'.

Capacity was one of those things that was rarely discussed outside of the business – but the capacity of crews' combined brains to operate an aeroplane and to think ahead was fundamental to operating her safely.

As a pilot, you're always thinking ahead. What will happen next? What *could* happen next? And if it does happen, what do I do? Your brain never rests.

In the days when passengers were allowed on to the flight deck, they'd likely see one of the pilots eating lunch while the other read a newspaper. Chances were they'd think: 'This is a piece of cake. I could bloody do that!'

But here's what's actually happening. There's a part of the pilot who's reading the paper that's thinking about what he's going to do if something goes wrong. On the VC10, I was subliminally given to doing this all the time. On Concorde, I'd be doing it, too, only I'd have half the time to react in the event something went wrong, because I'd be travelling at twice the speed.

There was so much more to consider – Concorde was a high/low, fast/slow aircraft, effectively four aeroplanes in one. If, because of a failure, you were forced to divert from your track to an *en route* alternate destination, all the operational characteristics of the aircraft at that moment changed – especially if you had to go subsonic, because your range would have reduced by 25 per cent – 33 per cent if the problem was an engine failure. If something went wrong and you had to descend from your 60,000 feet supersonic operating altitude to a subsonic operating altitude over the Atlantic, and you needed to do it very quickly, you'd be descending through other traffic patterns, having just been travelling faster than a bullet. And while you might know the tracks those other planes were flying, you wouldn't know where they were. So, thinking ahead, you needed to know about those tracks and the haste with which to get clearances from air traffic should something go catastrophically wrong. You also had to be aware, pretty much in real time, what the weather was doing at all your *en route* alternates. And you wouldn't have enough fuel – *ever* – to be able to go most of the way across the Atlantic and turn back. You had a tactical chart that showed your route across the ocean, with Ireland on the right and Newfoundland on the left. Concorde always flew the same track. Subsonic aircraft flew a different track with a different track structure each day because, flying westbound, you wanted to minimise the effect of the prevailing wind. Usually, there were five or six tracks, all parallel to each other, in a wide band around 120 miles apart (this has since dropped to around 60 miles). Inside each band there were six individual tracks, with each aircraft cleared to fly down one of these tracks at a certain height and speed, clearance being given along the lines of: 'Fly on Track Charlie at Flight Level Three-Five-Zero at Mach 0.80.' In those days, there was no radar or satellite navigation, so this was one of the ways ATC maintained separation. The band of tracks varied in its geographical location depending on the wind – and they were always assigned on the basis of minimising its effects. If there was no wind, you'd be plotted to fly the shortest distance, known as the 'great circle' – the shortest distance between two points on the globe. A great circle track is counterintuitive because you'd think a straight line would be a simple matter of sticking to a bearing on the compass. But bearings are measured against 'meridians', or lines of longitude, and these are not parallel – they converge as they come together at the poles.

The angle a great circle makes to the lines of longitude, thus, shifts constantly as you progress along the great circle.

Flying from LA or San Francisco to London, for example, your track changes along the route. Initially, you'll be tracking to the north-east, then east, and latterly, south-easterly. But, throughout, your great circle maintains its constant straight line.

Because, however, there was always a little wind, the tracks used by subsonic aircraft varied quite a lot – they could shift a long way north or south.

Concorde, needless to say, didn't use this system. We *always* flew the great circle.

At 60,000 feet Concorde faced wind, but the effect of a 50-knot headwind on an aircraft doing Mach 0.8 is significantly greater than its effect on an aircraft doing twice the speed of sound. As a result, it had been determined that it wasn't worth us chasing the wind on the day – that we would always fly a fixed track, this track being the great circle track system.

There were, in fact, three of these tracks: two westbound and one eastbound. These allowed us to maintain separation from other Concordes – not just BA's but Air France's – you'd otherwise meet coming the other way.

So, you were sitting up there on the great circle track and, below you, there could be any number of subsonic aircraft.

This was the kind of thing that a part of us had to compute all the time we were up there – it was what 'capacity' was all about. If our 'thought-bucket' was 100 per cent full, we wouldn't have had any capacity to deal with anything else; if it was empty, we'd simply switch off – no good either. The psychologists all said that somewhere between the two was an optimum, but it depended on the individual. Below 30 per cent you were in too low a state of mental arousal and above 80 per cent, you were in danger of overload.

So, Norman was right. Thanks to my 'non-handling' from the right-hand seat, I would get some operational instruction, but not so much that I'd not be able to take in the experience. Instead, I'd learn almost by a process of osmosis and absorption, what happened when you 'handled' the aircraft for real across the Atlantic.

Sit back and enjoy it; savour the experience, had been Norman's instruction.

My initial disappointment at not flying the take-offs and landings had been replaced by a feeling of excitement. This was absolutely what I intended to do.

15

Soon after getting airborne, Norman set course for the Bristol Channel and our acceleration point to Mach 1. It was a glorious morning, the sun shimmering off lakes and rivers below as we headed west to the sea. Because it was conical, our sonic boom went up, down and sideways, which meant we needed a certain width of water to accommodate it – at least 50 miles was planned – which is what the Severn Estuary and the Bristol Channel gave us.

Although we could begin our acceleration over land, by the time we went supersonic we needed to be over water, somewhere between the South Wales coast and the northern coastlines of Devon and Cornwall, the boom so weak by the time it reached them that it wouldn't be detectable, let alone rattle any teacups.

The Bristol Channel also happened to be a convenient point for us to begin our great circle track across the Atlantic to the east coast of the US, taking us conveniently close to the southern tip of Ireland where the circle began.

With Norman flying, as 'handling pilot', he would usually brief the passengers. But, today, and to my great delight, he asked me to do it.

There was no TV or movie system on Concorde – it had a great HiFi audio system, but that was all – so we, in effect, were the entertainment; and we had to put on a good show. For our customers, going supersonic was a big moment – for some, it might even be a unique, once-in-a-lifetime experience – and we didn't want it to disappoint. And this being my first time, I wanted to get it right.

So, I began to explain over the intercom what would be happening all the way up to and beyond the moment we passed through the sound barrier. On our way to the Bristol Channel, we'd be accelerating and ascending to our initial cruise altitude of 28,000 feet, I said, and at 95 per cent of the speed of sound: Mach 0.95.

A conventional aircraft, I continued, would be flying at 35,000 feet and Mach 0.8, but Concorde liked to travel quickly for the reasons I'd discovered in all those briefings: the drag on the airframe decreased the faster we went; Mach 0.95 being her most efficient subsonic speed, 28,000 feet being the point at which we were able to reach it; any higher and we'd have to slow down.

One of the *en route* points we put into the inertial naviga-tion system was the acceleration point – the INS giving us a countdown in miles-to-run to that point.

Norman, Graham and I had already carried out our tran-sonic checklist.

Everything was set up correctly to begin our acceleration to Mach 1.

As a crew, I explained to the passengers over the intercom, we would count down to the 'accel point'. Captain Todd, I said, would be handling the controls; I would be looking after the navigation and ATC clearances, and Graham Tul-lier, our flight engineer, would be looking after the systems and engines.

I gave the passengers a moment or two to absorb all this and then continued. 'As we approach the acceleration point, we'll open the engines up to full power and then switch on the reheats – just as we do for take-off.

'This time, however, we're going to switch them on in pairs, which means you may feel two slight nudges of acceleration as they come in. Fairly soon after that, you'll see us go past Mach 1, the speed of sound, on the indicators in front of you.'

Before signing off, I gave them one last reminder, as reassurance, that the slight jolts they'd be feeling would be the afterburners kicking in, not the sound barrier itself. I then told them I'd be back to brief them again in the run-up to Mach 1.

Norman turned to me. 'Good job.' Needless to say, as a good instructor, he'd been listening to my efforts through his headset. He glanced over his shoulder at Graham. 'Can we all pre-brief now?'

I looked up through the visor, where the blue of the sky met the deep blue-blackness of the stratosphere and the lower reaches of space. I nodded.

'When we get to zero-distance to the accel point, Graham's going to open up the engines to full power,' Norman said. 'As soon as they're at full power, he'll put the reheats on in pairs, inboards first, then the outboard.'

'I will then let the speed increase to V_{MO},' Norman said. V_{MO} was shorthand for 'Velocity – Maximum Operating'. Or, more simply, our top speed at the altitude we were at. It was indicated by a black-and-yellow needle on the airspeed indicator.

Concorde's instruments were analogue – what we referred to as 'clockwork', like the instruments you got in cars at that time – before 'glass cockpits', the computers and TV screens you get in modern airliners – and the V_{MO} indicator was no exception: it was a needle that rotated around a gauge.

As the handling pilot, from this point on, Norman needed

to fly Concorde at the maximum speed allowed because the faster she went, the more efficient she was. The designers had cut a small notch into the V_{MO} needle, into which you tucked the white airspeed needle. When the black-and-yellow needle moved, you needed to ensure that the white needle moved with it. This was because the higher you went, the max speed increased as well.

I glanced up through the visor again. Fairly soon thereafter, we'd be going through Mach 1.

There was a Mach meter in the cockpit that told you the speed of sound – or, rather, it told you your speed *relative* to the speed of sound.

We had a digital readout to tell us too. Neither, however, was the definitive indicator for breaking the barrier. This was an instrument called the 'vertical speed indicator', which was fed by pressure sensors on the airframe.

Some of these sensors measured static pressure – the ambient pressure – some measured dynamic pressure. As the sonic shock wave moved past the sensor plates from nose to tail it affected these sensors, particularly the vertical speed indicator, which, as its name suggests, measured rate of climb and descent in thousands of feet per minute, and wasn't an instrument we used very much.

But as the shock wave went past the sensors that fed it, the VSI went haywire – and this was our only definitive indication we'd actually gone through the sound barrier.

And it was handy because there was no other discernible effect.

'. . . and that, ladies and gentlemen, was us going through the speed of sound,' I announced over the intercom.

There was a story told of Sir George Edwards, the chief designer of Vickers-Armstrongs, the man who'd led the UK industrial effort behind Concorde, that he had once been upbraided by a lady of a certain age who had been provided with a once-in-a-lifetime Concorde experience. Towards the end of Concorde's test phase, she had been picked, along with ninety-nine other people – ordinary members of the British travelling public – to see what they made of the aircraft in a short hop that took them well through Mach 1. The lady in question had never flown before. Sir George, thus, wearing his overcoat and trademark pork-pie hat, was keen to know what she had made of it when he'd greeted her and the other passengers off the flight at Filton.

'It was lovely,' the lady had said, 'but I was a bit disappointed, truth be told.'

'Oh? What by? When?'

'When we went supersonic.'

'What were you expecting, Madam?' Sir George enquired lightly.

'Have you seen that film?' she asked him. 'The one with . . . *whassisname*, Sir Ralph, er—?'

'Richardson?' Sir George prompted politely.

'That's him,' she said.

'You're referring, I take it, to the David Lean film, *The Sound Barrier*,' he replied.

Which indeed she was. In the film, which had been released in 1952, at a time when British test pilots were still being killed as they probed the speed of sound, the 'trick' to the dreadful, bone-jarring 'buffeting' that was depicted as the aircraft went through the sound barrier, was a fictitious conceit: to get the prototype through the 'barrier' safely, it was discovered, right at the end, the hero-pilot had to reverse the controls. The buffeting that had shaken the plane almost to bits, though, had made a lasting impression on the public.

'I was expecting it to be just like in the film,' Mrs Bloggs said, a little disappointed, apparently, that it hadn't been.

Sir George, the story had it, had given no hint of any temptation to roll his eyes. 'That, Madam,' he said, quietly, 'is the clever bit.'

And it was. So clever, in fact, that – just as everyone had forewarned me – it had been a sensation-free experience. But emotion-free? Not at all. It had been like my first solo. I felt elated.

Thirty years, almost to the day, that Chuck Yeager – the legendary American test pilot – had broken the sound barrier for real, I had just done the same thing. And, like some of the people in the cabin behind me, I was in shirtsleeves. Yeager had been in a 'pressure suit' and his rocket-powered aircraft had to be dropped from its 'Mother Ship' because it couldn't take off under its own power. I wasn't sure too many of our passengers would have put up with that.

The headiness of it all almost got to me – especially when I thought about a fact that would have been as astounding to Yeager as it would've been to the cast and crew of *The Sound Barrier*: in just a few hours we'd be landing at Washington Dulles – an hour and thirty minutes *before* we departed Heathrow.

16

Capacity . . .

The first place my mind slewed to as we went transonic were the ramps regulating airflow to the engines during our supersonic regime. The ramps didn't start operating until Mach 1.3. This was the next critical 'gate' in our flight envelope, and it was rapidly approaching as we began to pick up speed out over the Bristol Channel towards the southern tip of Ireland.

The ramps didn't just reduce the 'throat' of the intake, they generated a whole pattern of shock waves from the lip of the intake ramps to the face of the engine.

As air transited this series of shock waves, it slowed and compressed.

Slowing the air, as my briefings reminded me, reduced its velocity to a speed the engines could handle, which was less than 500 mph. Compressing the air was something like the function performed by my car's turbocharger.

The task before the Olympuses was to get this air back up to speed again as rapidly as possible as it exited the engine, not as easy as it sounded given that it had to transit fourteen stages within the engine core – six at low pressure, eight at high pressure – akin a bit to a boat making its way through a chain of lock gates with each gate rotating at a fantastic rate of speed.

At the back of the engine were two sets of nozzles – primary and secondary. The primary was a petal-like duct at the end of the jet-pipe; the secondary a set of clam-shell doors, which extended aft of the jet-pipe.

As well as activating on landing to give us reverse-thrust, the clam-shell doors – also known as 'buckets' – could 'modulate'; that is, they could work with the opening/closing petals of the primary nozzle to balance the pressures between the two shafts. It was this act of magic, unique to Concorde, that allowed it to operate so efficiently at high speeds.

And it was so cunningly designed that the compression and expansion action – this ram-in, ram-out process at the front and back of the engine – generated more than 60 per cent of our thrust at Mach 2. The other 40 or so per cent came from the core of the engine itself.

Developing the engine and the intake system had been a uniquely British contribution to the Concorde effort – a brilliant piece of work at that. It was also so sensitive that the Brits had had to design the computers that regulated the millimetre-precise positioning of the ramps. Knowledge that the ramps would work as advertised was one of our key concerns during a flight.

I glanced down. The Mach meter was nudging Mach 1.3.

Were the ramps not to work, there were a couple of things we could try.

To begin with, control would be automatically assigned to another computer – there were two per intake.

If that didn't work, we'd try diverting hydraulic power from another system in the hope of giving the ramp the requisite muscle to make it function.

If neither of those did the trick, we were stuffed: without full operation of the ramps, we couldn't go supersonic – and no supersonic, no transatlantic flight. If we dropped to subsonic speed, our range instantly reduced by 25 per cent.

As counterintuitive as that sounded, it was a fact. Our fuel burn might be less, but when you're travelling supersonically, you're going twice as far, and your consumption when you flew supersonically, although more, wasn't twice as much, so it was just a 'function of the math'.

Another glance at the control panel. Mach 1.3 and the ramps *had* kicked in, modulating exactly as they were supposed to.

My mind jumped to the next portion of the flight as, behind me, Graham Tullier, the flight engineer, prepared to oversee the transfer of fuel aft.

Balancing the aircraft – as had been rammed into me at those first briefings – was critical to the way she flew; *de facto* to her safe operation.

Years later, in the wake of the crash, as the scope of our inquiry widened, we began to put C of G and the fuel transfer operation under the spotlight as a possible contributory cause – and, as it turned out, we were right to do so.

Today, though, thoughts of a disaster were far from my mind but, due to its criticality, as pilots, Norman and I monitored the process all the same.

At take-off, the C of G had been set at 54 per cent. Now, to re-trim the aircraft, Graham needed to move the C of G back to 59 per cent – by around 6 feet.

This necessitated the in-flight movement of as much as 20 tonnes of fuel, a calculation Graham was able to make by virtue of three 'knowns' that had been inputted into the computer via the load sheet at Heathrow: total fuel on board, the aircraft's ZFW 'zero fuel weight'– its weight with everything on board *apart from* the fuel – and its ZFCG: its C of G position with zero fuel.

Knowing these figures told him not only how much fuel

we needed to burn or transfer to set the C of G correctly for take-off, but also, now we were poised to move into the cruise portion of the flight, how much to transfer into Tank 11 beneath the fin to trim us for supersonic flight.

The centre section of Graham's panel was the fuel-control section. On it was a computer, into which he had inputted the ZFW and ZFCG figures.

The computer, which knew how much fuel was left in each tank, was able to calculate at any given moment where the aircraft's C of G was meant to be.

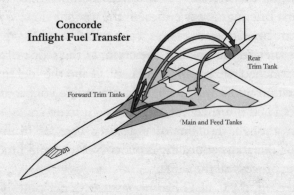

Concorde Inflight Fuel Transfer

Rear Trim Tank

Forward Trim Tanks

Main and Feed Tanks

The fuel transfer system itself was a semi-automatic process. Graham told the system how much fuel needed to be transferred and primed counters in Tanks 9, 10 and 11 to shut the process down as soon as the right amount had been moved across.

He now threw the 'Trim Transfer Auto Master' switch to the rearward position – and the transfer began, a myriad of pumps and valves coming to life in accordance with the transfer schedule.

As this was happening, Norman and I were getting a real-time read-out of the C of G position.

The aircraft now really began to accelerate as it climbed towards 41,000 feet *en route* to our operating ceiling of around 58,000 feet. The VC10 had often flown at 41,000 feet – it was the *speed* at which we were flying that was now starting to become really noticeable. I glanced down at the navigation panel: 600 miles per hour, the cruise speed of a VC10, was the speed I was used to – a mile every six seconds. Here, now, as we began to move almost twice as fast, I tried to imagine what twice as quick looked like on the ground: a mile every *three* seconds. As I glanced out the window, it didn't appear that much different – until, several thousand feet below, I spotted a 747. For a brief moment, even though it was pointing in the same direction as us, it seemed to be flying backwards. That, I suddenly realised, was a measure of our speed.

The sky was beginning to darken. I glanced back over my shoulder. The 747 had vanished into the glare. Far below, as we continued to climb, the Welsh coastline receded into the mist of an early autumnal evening. Ahead, to the right, the Irish coast loomed out of the haze. The Mach meter began inching towards Mach 1.7. Behind me, Graham announced that he was switching the reheats off. Norman and I concurred. As the afterburners winked out, I half expected to be thrown against my straps, but no such thing happened.

We continued to climb, up and up, on 'dry' power alone, into the darkening sky.

17

The engines were now doing their thing – and Concorde began to perform as no other aircraft did: climbing and accelerating from Mach 1.7 to Mach 2 using full power without afterburner.

And then I really did begin to feel the darkening of the sky, as the atmosphere thinned and the light-scattering properties of the air diminished.

Out to my side, I saw the curvature of the Earth – I'd been told what to expect but had no comprehension of how stunningly beautiful it would be. As we continued to edge towards the inky blackness of space, the Mach meter now nudging Mach 2, the horizon bowed a little more, and for a second I was that boy on the beach again, gazing at the azure beyond the silver specks – the Viscounts, Ambassadors and DC-3s making their way to the Continent.

Norman looked at me.

'Thank you,' I said.

'What for?'

'For insisting I fly this trip as co-pilot.' I took my camera out – a little Olympus Trip – and snapped several shots out the front and side windows.

'It's quite something, isn't it?'

'Yes,' I replied. 'It is.'

'We had an archbishop on the flight deck once,' Norman told me. 'Said it made him feel closer to God,' he chuckled to himself.

I smiled, but the man of the cloth had been right. I'd heard astronauts speak of the awe they'd felt from gazing down on the Earth from space. The elements were all here: the darkening sky, the blackness above, the blue below, the curvature of the horizon ahead, a realisation that there was no one above us; that we were it. And, apart from a few satellites, nothing aside from us that was even man-made. We were at 58,000 feet, 2,000 feet short of Concorde's ceiling of 60,000 feet. The height we flew was a function of the temperature, which, counterintuitively, was colder over the equator at this height than it was over the Pole. On our great circle track to Washington, it wasn't cold enough to go to 60,000, but when we went to Barbados, we did.

I glanced down at the nav panel. Mach 2 – and those mileage numbers were clicking down a darn sight quicker: every two-and-a-half seconds another one.

'Makes you think, doesn't it?' I said to no one in particular.

'What does?' Norman said.

'Being up here. I used to live a mile-and-a-half away from my school. It took me half an hour to walk there and half an hour to walk back. And now I'm covering the same distance in a few seconds. Hard to get your head around.'

I looked down. There was still no sensation of speed. The regular traffic was over 20,000 feet below us. Occasionally, I saw a contrail with a 747 or a DC-10 at the end of it or heard some of their chatter over the radio. It was hard to decouple from the majestic solitude and the dreaminess that came with it, but there was business to look after.

My main point of focus now was the functioning of the aircraft – was she operating properly? For all her thoroughbred qualities, Concorde was inherently reliable. But still. My brief moment of spiritual transcendence over, the voice in my head kicked in again – the voice that said: *capacity, Mike.*

What would happen now – right this second – if an engine were to fail?

What was the engine failure drill?

What were all the things I had to remember on that yellow checklist?
While the aeroplane was physically capable of flying super-sonically with an engine out, what would I *actually* do? We'd be down to Mach 1.3 and into the high-drag region . . .

First things first: to be more fuel-efficient, we'd need to decelerate to subsonic speed and descend.

Next: where were all those 747s and DC-10s? Where was the track system?

Right below us. OK. So, how would I get myself down through the track system, amongst aircraft whose position I was uncertain of? Concorde's radar didn't show aircraft, just significant weather, and by now we were well beyond air traffic's capacity to advise.

We're heavy, and if an engine was out, I wouldn't be able to maintain three-seven-zero as a flight level. I'd have to go lower, descend past the traffic in the track system. I'd have to position us so we were halfway between the tracks, which were 120 miles apart.

And then, where would we go? At this point, we wouldn't have enough fuel to continue to Washington. We would be able to turn back – the little flags on my chart told me we'd have enough fuel for Shannon or Prestwick, but not for London; we were past that point, which was always a moment, because were anything to go wrong and we couldn't go on, Heathrow was the place we'd want to make for – Heathrow, where we had engineers who could fix most problems or, failing that, a replacement aircraft on standby to get our VIP customers to where they needed to go.

This was how the mind of a pilot worked – it was certainly how my mind worked – and especially over the Atlantic.

We always started on the assumption of an engine failure because (short of an engine *fire*, which, after it was extinguished, elicited the same response) it was the most critical failure we were ever expected to encounter.

The picture changed constantly because our position relative to the tracks was always changing, the amount of fuel we had on board changed, and our ability to turn back or continue changed as a result.

Once past the point of no return, you had no option but to press on.

When we reached that point a whole new set of questions kicked in.

Would we be able to make Newfoundland? Or, faced with a choice of alternates, which airport would be best?

Gander in Newfoundland would be closest, but what about the customers? How likely would they be able to get a flight from Gander to D.C.?

I had enough fuel to get to Bangor, Maine. Bangor was busier than Gander – lots of flights to points all over the US – so Bangor, at this point, would be better . . .

And since I needed to fly by Gander to get to Bangor, if the situation got any worse, I'd still have Gander as another alternate if we had a second problem . . .

All of this I shared with Norman and Graham. They, in turn, discussed my choices, until, via a process of elimination, we had a set of options we all agreed upon. But then, because of the dynamic nature of our situation, within a few minutes, the situation had changed again – and the plan was updated.

Meantime, as the non-handling pilot, I was expected to be fully *au fait* with the weather in all of the alternates we'd just discussed. One of the things I'd also have to do should we have an emergency – a priority only when we'd sorted the problem – was getting in touch with BA, which wasn't as easy as it sounded. We couldn't talk on the VHF radio, because base by now was out of range, so we'd have to use our HF radios instead – the frequency selected being a function of distance, the time of day and the atmospheric conditions; and often we'd try two or three different frequencies before getting through.

But assuming, of course, nothing went wrong, and with our destination now less than a couple of hours away, I had to think about contingencies at our destination: Washington Dulles. What time would we arrive there? What would the weather be like? Were there any air traffic delays that might affect us? How much fuel would we have left when we got there?

I sat back. For the moment, I had all the bases covered. But because the capacity process never stopped, my mind was never far from computing some new solution.

My God, I thought, *I'm 28, still cautious about driving my car, and they've let me have a go at all this.*

18

I was having so much of a ball, in fact, that I suddenly realised it was coming to an end all too quickly. The time was approaching when we had to slow the beast down and descend into normal traffic for the final stage of the flight.

Today we'd been routed on the great circle track taking us into the US over Nantucket Island, 100 miles to the southeast of Boston. We would, however, have to be subsonic long before we got there because, as the aeroplane slowed, travelling at twice the speed of sound, that shock cone of ours was travelling twice the speed of sound as well. When we reduced power, bringing the aircraft down from Mach 2 to Mach 1.9, the plane itself might have slowed, but the cone we'd made a few seconds earlier was still travelling at Mach 2 – we were throwing the boom forward. As we reduced to Mach 1.8, our boom was still being thrown forward at Mach 1.9. And so on and so on. The long and short of this process of slowing – with our concertinaing, forward-thrown boom – meant we had to get subsonic at least 50 miles before we reached the coast.

We also found out after some initial experience that, under certain conditions, the boom would bounce off the upper atmosphere – that cone went up as well as down and to the sides – meaning it went even *further* than we'd originally anticipated. Knowing when we'd have these peculiar conditions required us to be subsonic even earlier than normal – maybe as much as a hundred miles before we hit the coast. All of which meant we needed to sort out and agree amongst the three of us exactly when to begin the deceleration. We were making for a fixed point – a radio beacon on Nantucket Island.

Using a set of tables in the flying manual – a document called the Normal Check List – we were able to work out exactly where the 'decel point' was.

It took something in the region of 105 nm, nautical miles, to slow from Mach 2 to subsonic. Add to that the 50 nm to compensate for our thrown-forward boom and we were looking at a point 155 nm from landfall.

These kinds of decisions, however, had always to be agreed between us. It was never something the handling pilot worked out on his own.

Graham and I were busy calculating the decel point independently.

When we were done, we cross-checked.

Having agreed what the distance would be, we then fine-tuned the parameters.

Generally, upon reaching Mach 2 we'd ascended to a height of around 50,000 feet. As the aircraft by then had burnt fuel and become lighter, we'd then climbed gradually over the Atlantic to a height of 58,000 feet. Air traffic allowed us to remain in a 50,000–60,000 feet band – where we were in that band depended on the instantaneous weight of the aircraft and the outside temperature. Whilst generally climbing, we would drift up and down in the band to some extent. We used a clever autopilot function called 'Max Cruise', which held Mach 2, or the maximum temperature on the nose, as conditions changed.

When we reached our decel point, we needed to be at a constant altitude. So, Norman selected the autopilot into its 'Alt Hold' function.

The decel point was critical – we had to hit it bang-on – so we peeled our eyes as we stared at the counter that took us down to the 155 nm mark.

I popped out the auto-throttles and counted: '*Ten, nine, eight . . .*'

When I got to 'one', Graham leaned between us and pulled back on the throttles. This, too, required focus. If he reduced power by too much (though, it had to be said, by quite a lot too much), the engines could 'surge' – there would have been more air rushing into them than they'd be capable of handling. Gradual reduction was better, too, the maintenance people told us, for engine life.

Graham brought the throttles to their interim setting of 18 degrees TLA or 'Throttle Lever Angle' – just under halfway.

Whilst maintaining 58,000 feet, we came out of Mach 2 and began gradually to reduce our Mach number, and our IAS indicated airspeed: 520 knots back to 350 knots.

At this point, we needed to transition from holding *altitude*, which is what we'd switched to as we approached the decel point, to holding *speed* – the 'IAS Hold' function. We were at about Mach 1.6 at this stage and then, at Mach 1.5, Graham closed the throttles further, to 24 degrees TLA – about two-thirds back.

As soon as we got to Mach 1.3, he pulled the throttles closed to idle and we descended to our pre-set subsonic cruise level, which, for the first time in several hours, was the same as the altitude at which 'normal' aircraft flew; the biggest difference between them and us – we were still going significantly faster.

Our main concern at this point was alerting air traffic to our significantly higher speed – one of the reasons we had the callsign 'Speedbird Concorde', so as to differentiate us to controllers from conventional aircraft in BA's fleet.

To begin with, our subsonic colleagues thought this was us showing off, another instance of 'Concorde elitism', but the real reason was to alert air traffic to just how fast we were going relative to everything else. Somehow, between now and landing, we had to integrate with all that slower traffic.

Most of the gee-whizz stuff of the flight was now over. I was entering into a routine that was familiar to me from all my simulator training. I'd also had a chance to practise real-world approaches and landings during my base-training at Brize. But we were still travelling pretty fast and in addition to my now having to focus on tasks that I was having to carry out in real-time so as to support Norman, I was also doing the 'capacity thing': leaving a part of my mind free to compute a bunch of what-ifs when we got to Washington Dulles.

What if the aeroplane ahead of us burst a tyre on landing? What if we couldn't land on one of its other runways? Because we'd been travelling twice as fast at this point than I had in any aircraft I'd flown before, the pace at which this was having to happen might have had my head spinning had it not been for the quietly reassuring presence of Norman and Graham. *What next?*

Lowering the visor and nose.

We did this as we decelerated through 270 knots to 250 knots.

At 270 knots, Norman called: 'Visor down, nose down to five.'

The visor only had two positions – up or down. The nose had three positions: up, the high-speed position, which meant in excess of 250 knots – the position it had to be in for supersonic flight and the position we left it in on the ground (partly to keep the rain out, but mainly because it looked prettier that way); the other two were halfway down and fully down.

Halfway down was 5 degrees; fully down was 12.5.

The lower it was, the more drag it produced. We only needed 12.5 degrees when we were really nose-up on approach.

The 5-degree position was for take-off and for intermediate approach – at or below 250 knots. We selected it fully down just after we put the gear down and were on 'slow', final approach.

Not that 190 knots to 800 feet and then around 162 knots from 500 feet to touchdown was slow to this ex-VC10 pilot. It felt as if we were approaching our destination in a rocket ship.

Upon Norman's command – 'visor down, nose to five' – I placed my hand on the lever on the instrument panel to the left of my control column. Graham, as the flight engineer, checked and verified I had the correct lever. Norman – experienced pilot that he was – did too. We were all human – a momentary lapse might have seen me lower the undercarriage instead. And, besides, it was a mark of the discipline that had been instilled in us from the get-go – cross-checking everything that each of us did. This included all the procedures we now needed to carry out prior to landing: ensuring fuel had been correctly transferred to the correct tanks; that we were agreed on our approach speeds; that the centre of gravity had been moved forward to its correct position for landing; and that the cabin had also been secured for landing.

A conventional aeroplane on final approach would be doing 120–130 knots; our final approach after a transatlantic crossing was more like 162.

This now brought us to another unique Concorde phenomenon, one related to its 'minimum drag speed'.

On final approach, with our nose up, the drag on the big delta wing was considerable.

At low landing weights, our minimum drag speed was about 270 knots. If we dropped our speed lower, we'd produce more drag, which meant we needed more power to overcome it – just to keep flying. The highly counterintuitive corollary to this was that to go slower, the more power we needed from the Olympuses – and *increasingly* more. This, inevitably, produced a lot of noise.

To cut the noise down meant – again, seemingly in the face of all logic – flying as fast as we could for as long as we could – and then implementing a technique on short finals that we called the 'reduced noise approach'. This required us to fly at 190 knots for as long as possible before dropping to our 162-knot final approach speed.

But when to reduce?

Riding what we called the 'back of the drag curve' was demanding because the slower we went, the more the aeroplane became unstable. With a car, run it downhill and the faster it goes. With Concorde, it was the opposite: unless you put more power on, the rate of deceleration increased. To help crew handle this situation, the designers had to invent a sophisticated automatic throttle system – akin to an autopilot. It comprised a window with a read-out and a dial. You dialled in the speed you wanted to go and the aircraft maintained it all the way to touchdown. It was so good at doing this – at riding that drag curve – that there was no occasion in which we were allowed to fly approaches *without* the autothrottle; not even in practice. As with so many other critical features of the aircraft, the auto-throttle had a second as back-up. If one failed, the other automatically kicked in. Coming in, we selected the final speed we needed via one of two modes: 'IAS HOLD' or 'IAS ACQ'. The former stood for 'indicated airspeed hold' – like cruise control on a car, you pressed the button and the aircraft would hold the speed you were flying at.

With IAS ACQ – for 'acquire' – you told the computer what speed you wanted to fly. Press the button and the aircraft automatically accelerated, or decelerated, to that speed and then held it.

Norman selected IAS HOLD at 190 knots.

After pre-agreeing what we'd do next, I then dialled our touchdown speed into IAS ACQ: 162 knots.

A bar in the window indicated the number was in the system, but not yet active – exactly what I wanted to see. As soon as we pressed IAS ACQ, the aircraft would drop to a 162-knot approach speed. It was thus we were able to delegate to the aircraft the mechanics of a reduced noise approach.

Watching the throttles move as they went through this procedure on their own was quite eerie. When we hit IAS ACQ, they motored back all on their own to reduce power. Meantime, we had our attitude to think about.

I looked at the airspeed indicator. At 190 knots, we were already at a steep nose-up attitude. At 162 knots, it needed to increase. In order to maintain our glideslope, Norman was adjusting the controls all the time to maintain the correct angle.

Somewhere amidst my thinking, I was listening, too, to the pitch of the engines, which were spooling down, reducing our noise.

I glanced out the window and saw the lush green countryside outside of the US capital.

The weather, we'd been told, was good, with clear

visibility all the way down to Dulles's long runways. The 800 feet marker was a function of several things. The one thing a pilot wanted on landing was stability. You didn't want things changing close to touchdown – and this was what the 800 feet marker gave us. Now we were going to deliberately upset all that stability and change lots of parameters all at once. Our aim was to be reset and stable by 500 feet. On a conventional aircraft that would be bonkers – and prohibited.

At 800 feet, the power would come back on to bring us down to 162 knots – that Concorde phenomenon that produced a change in noise as we slowed. Whilst jarring compared to a conventional aircraft, we'd likely then be within the airport boundary, allowing us to claim our 'reduced noise approach' as advertised.

Just before we got to about 500 feet, the speed had fallen to just above our target 162 knots and the throttles did their own thing again as they put more power on – more power, and hence more noise, than we'd had at 190 knots. Initially, that seemed a bit odd for a 'reduced noise' approach, but the key was to realise that, between 800 and 500 feet, the power was right back and noise was significantly reduced. This period corresponded to being both close to the ground and close to the airport, so folks who lived about 2 to 3 miles out from touchdown got a lot less disruption. Much closer than that, and you were inside the airport boundary – so, no people to disturb.

And it was good for those further out as well, because the power needed at 190 knots was markedly less than at 162 knots. None of this was to say that Concorde was quiet, only that we did our very best to reduce the impact on local ears.

As the co-pilot, I was deeply involved with Graham in monitoring all of this.

Norman, as the handling pilot, had his hands full maintaining our attitude on approach, although he kept his right hand lightly on the throttle-levers, their minute adjustments telling him what the engines were doing.

He left it to Graham and me to check that the IAS ACQ command had fed into the system as it was supposed to. If, by 500 feet, the aircraft hadn't reset to our 162-knot final approach speed – if, for some reason, there was a problem with the automatic throttle, and it didn't pile the power back on – then we'd be going slowly. Not dangerously slowly, but slowly. The 500 feet marker, thus, gave us enough height to take remedial action if we needed to.

The undercarriage was already down – I had selected it around 9 miles out. This was further than a conventional aircraft – approximately twice as far. Lowering the undercarriage produces drag on any aircraft – more drag, means more power, which means more noise, which is why all pilots seek to lower the wheels as late as possible.

Once again, with Concorde, we found that lowering the wheels didn't impact the environment in the way that it did with other aircraft. Because of that massive wing – which, by now, was starting with our nose-up attitude to ride the drag curve – the dropping of our undercarriage into the airflow didn't make a whole lot of difference to the overall drag.

The reason we selected the undercarriage down as far out as we did was to leave enough time to lower it by other means. The inherent reliability of the undercarriage meant this was extremely unlikely, but in amongst my bag of 'what-ifs', it remained a possibility.

On a conventional aircraft, if you had a problem with the wheels, you simply went to a place where you could get the problem sorted and then re-entered the landing circuit.

With Concorde, you wanted to avoid going around, because going around made a lot of noise and burnt a lot of fuel.

Second, we were selling time and speed to our customers, who expected to arrive on time. Shooting off somewhere to get the landing gear fixed would gobble up 15–30 minutes of precious time, in addition to fuel.

Concorde would arrive at a destination often with not a huge amount of fuel over and above what it had to have.

On some operations, we were really tight. There was a minimum amount we had to have to 'hold for 30 minutes and then divert', plus a 'contingency' – and we'd always ensure we had that.

But there were some occasions when we were right on the limits, which was why we activated the undercarriage further out. The last thing we wanted was to 'Go Around'.

There were four ways to lower the gear, one more method than a conventional aircraft, so not unusual for the Beast. If 'conventional' had two, Concorde had to have three. If conventional had three, we had to have four – not a bad principle for a unique aircraft that flew where no others did.

Using the first 'normal system', 9 miles out, gave us time to work on Method Two if needed.

This was a hydro-mechanical and manual sequencing and timing system operated by a separate lever at the back of the centre pedestal instrument panel, between Norman and me.

The third and fourth standby systems were more intricate – to be used only as a last resort.

Then we would have to abandon the approach and go and get the problem sorted. That would involve a trip back into the passenger cabin and raising a few floor panels.

The 'free-fall controls' were located above the nose gear

and the main-gear up-locks. Removing the panels and getting to work on the locks would certainly have raised an eyebrow or two amongst the passengers, as Method Three entailed manually releasing the locks and using the slipstream to help the gear legs to descend under their own weight.

If that didn't work, then it was Method Four: releasing the lock and allowing the nosewheel to fall into the doors, pushing them open, then attaching a pneumatic air supply to the top of the telescopic struts and literally blowing the main gear legs out into the slipstream.

The slipstream then caught the nose gear, pushing it into down-lock.

According to the manual, a practised hand would be able to achieve three 'greens' on the landing-gear control panel in two minutes, 45 seconds.

Naturally, it was never expected to come to this and, except in training, it never did.

But it was a measure of all the redundancy that had been built into the aircraft that it was there if we needed it.

Coming in to land, my job was to monitor everything Norman was doing. As he raised the nose and flew down the glideslope, everything he'd done so far was within a whisker of everything I'd gleaned from the sim and the manual.

Because our attitude was so nose-up, it was challenging during this portion of the flight to judge height above ground. To get around this, we relied on a radio-altimeter positioned on the aircraft's belly to transmit a radio signal to the ground, pick up its reflection and interpret the returned signal in the form of height – very accurately, too; to within a few inches. The rad-alt was calibrated to give us the height of the main wheels above ground. It was down to me to monitor the readings and to Graham to make our height calls.

Moving past 2,500 feet, he announced, 'Radio altimeter active' when its needle appeared from behind a screen on the instrument.

Fifteen hundred feet later, he told us we were down to 1,000 feet.

That was the cue for me to check that we had 'five greens': one green light for each of the two main wheels, one for the nosewheel, one for the tail wheel and one for the nose itself. All had to be fully down and locked to give five greens.

It was also the moment at which I inputted into the auto-pilot and flight director the altitude we had to climb to should we need to go around. I called, 'Five greens – go around altitude set'.

Graham's next call was at 800 feet, at which point I was ready for Norman to call to me to select 'IAS ACQ 162', which he duly did. I pressed the button and monitored that the action had been implemented as I'd requested.

Three hundred feet later, I was able to report that the speed reduction that had been set into the auto-throttle had been completed successfully by announcing 'stable'. Graham called each 100 feet block below that down to our 'decision height' – 200 feet – the lowest we could descend when manually flying without a categoric assurance we were able to land safely. The predominant reason for a go-around at this height would be if the runway wasn't clear or the aircraft wasn't stable enough.

But outside, everything was looking good.

At Graham's '300' call I said, 'One Hundred Above,' and then, 'Decide' when we reached our 'decision height'. Norman then had to commit to whatever he was going to do; to which there were only two possible responses: 'land' or 'go around'.

Norman's firmly stated 'Land' signified that he was happy to commit.

Outside, beyond our steep, nose-up attitude, I had a clear view of Dulles's pale, concrete main runways and, in the distance, the distinctive white sloping roof of its once-*avant garde* terminal building, bright in the evening sun.

From 50 feet down, Graham called our height in smaller increments: '50, 40, 30, 20, 15 . . .'

At 15 feet, his final call to us, Norman disconnected the automatic throttles, closed the throttles and flared – not, for all the reasons I'd learnt on the simulator and at Brize – the reasons that conventional aircraft flared – but to overcome that unique Concorde trait: to stop the nose pitching down.

The pitch-down happened as soon as Norman closed the throttles – but by counteracting with a little backward pressure on the control column, Norman rode the flare down to the ground.

Next, I monitored that he had taken his hand off the main part of the throttle and moved it towards the reverse levers. These were additional levers at the front of the throttle you grabbed and pulled back to put the engines into reverse. The only point at which this could happen was when they were set to 'idle' (the Olympuses didn't *actually* go into 'reverse' – the clam-shell buckets closed aft of the exhaust nozzles to throw their thrust forward).

There was the merest of bumps as the main wheels greased the concrete. Norman selected 'idle reverse' to get the buckets into place.

It took a few more seconds for Norman to 'land' the nosewheel, call 'stick forward' and activate reverse thrust. If we didn't push the stick (or control column) forward the aircraft could rear up like a stallion when reverse power was applied, so I confirmed the command by repeating 'stick forward'. Next, I called out our ground speed as we slowed on the runway.

At 100 mph, Norman deselected reverse power by putting the outer engine pairs to idle, whilst braking at the same time. At 75, he did the same with the inner reversers. Because his pedals and mine were linked, I was able to feel the pressure he applied. I took my feet off just before we turned off the runway. We then raised the nose to 'five'.

While taxying to the stand, Norman and I began to 'tidy up', while Graham began pumping fuel forward again. A minimum of 4 tonnes had to be transferred in this way in order to put weight onto the nosewheel. This not only helped for steering purposes but ensured the aircraft didn't end up on its tail when the passengers disembarked from the front. This was unlikely, but it could happen – no crew wanted to be the first to suffer such an ignominy.

As we drew towards the stand, we lined up with the guidance system that helped us to park. It was a bit like going into a car wash. Ahead was a big square box. In the middle was what looked like a letterbox with painted yellow lines above and below it. Inside the letterbox was a fluorescent yellow light. All we had to do was manoeuvre the aircraft so that the light joined with the two painted lines and, *bingo*, that put us bang on the centreline. How far forward you went varied from plane to plane. Sometimes an intelligent sensor at the stand, pre-set with the type of aircraft you were, signalled for you to 'stop'; sometimes there was a board with a line that indicated 'stop points' for each aircraft type.

Ahead, an illuminated sign commanded us: STOP.

'How was that for you?' Norman said, as he shut down the engines.

I was still on cloud nine.

The veteran pilot looked at me and smiled. 'Nothing left to do now, dear boy, except sit back and listen for the applause.'

19

Not only was there a fascination for the aircraft amongst the crews who flew Concorde, there was a fascination amongst the people who flew *on* her, too.

Throughout the history of the aeroplane, right up to the point when we announced she would be retired, the passenger profile didn't change. The people we flew were 80 per cent business, 10 per cent rich and famous, 5 per cent in the sports, leisure and entertainment industries. The final 5 per cent were what we called the 'trip-of-a-lifetime' demographic.

Of the 80 per cent business, 80 per cent of them were repeat customers – which meant people who had flown on her five times or more.

Early on in the game, we identified an interesting phenomenon: that the number of people travelling tracked almost exactly the performance of the FTSE and the Dow. The more astute members of BA's staff noticed that our actual passenger profile was slightly *in advance* of those two indices – that, if our numbers suddenly dropped, the economy was due for a downturn and vice versa. This was so bankable from a stats point of view that some people even made money out of it.

In those days, if a passenger asked to come up onto the flight deck, nine times out of ten we'd say yes. Back then, the door of the flight deck was almost always open – and not just to keep us close to our customers. The aircraft got hot. The area with the best air-conditioning was the flight deck.

The hottest area was the forward galley. The fuselage heated because we were travelling so fast. The friction and compression of moving through the air at twice the speed of sound would generate temperatures on the nose of up to 127 degrees Celsius. The temperature on the rest of the structure was *only* around the boiling point of water: 100 degrees Celsius. The result was that the airframe would stretch by as much as ten inches in flight and the inside of the cabin would get warm – parts of it very warm indeed. The air-conditioning lavished on the flight deck wasn't for us poor, delicate pilots – it was for all the instruments and computers. But a happy by-product was that it kept us cool; and we felt it only right that we shared some of our nice, cold air with our colleagues in the galley – hence the open door.

This, in turn, encouraged people to ask if they could visit the flight deck. After a while flying Concorde, you would have thought that for our repeat customers the novelty of flying the aircraft would have worn off. But it never did. It would be an unusual transatlantic trip if at least ten of our passengers didn't come up to the flight deck. Often, it was more. Later, when we began to use Concorde for charter flights, it would be all one hundred. Everyone just wanted to have a look and to talk to us about it all. We loved it, partly because we got to meet the customers, partly because we could show off our shiny new toy and partly because it was all good, positive PR for BA.

The passengers ranged from the experienced traveller to the trip-of-a-lifetime couple who couldn't quite believe where they were.

Occasionally, we'd invite individuals to come up for the take-off or landing. That could be at the behest of the company – we'd get a message, 'extend all courtesies', which was code for 'if you can invite them on to the flight deck, they'd be delighted'. Sometimes, we were asked from on high because the passenger was particularly important to the airline; other times it was somebody who was known to the crew or simply a request from a repeat passenger we'd come to know. Or it was someone who asked nicely – or maybe a celebrity. Or a new Concorde cabin crew member on the first flight after their training. Bottom line for us, it didn't matter. Considering how much excitement I got – throughout my career – just from piloting her, I could only imagine the impact it had on our passengers, whether they were with us during take-off, landing or just cruising along at twice the speed of sound. And to the extent that it was practical, we tried to explain everything we were doing. For take-offs and landings, they were fully strapped into the fourth seat. If we were in the cruise, they'd simply stand at the back. Occasionally, I'd turn around just to catch the look on their faces. The sharing of our experience with our passengers was one of the joys of being Concorde flight crew.

Because the cabin window was quite small, you were relatively restricted in what you could see compared to the view you got from the flight deck, where you got a far better view of the darkening sky and the curvature of the Earth and the occasional, fleeting view of a backward-flying 747.

Despite the visor being up in supersonic flight, from 58,000 feet you could see a quarter of a million square miles of the Earth's surface.

Heading the other way towards London, the shadow of the Earth was a very distinct line in the sky that would race towards us as it got dark. In technical circles, it's known as The Terminator – the twilight zone that divides the daylit side of the Earth from the gathering night.

But an even more awe-inspiring sight awaited those who shared the flight deck with us on spring and autumn flights heading west.

We'd take off around 7 p.m., by when it would be dark. By the time we got up to 1,350 mph, we were travelling faster than the Earth rotated – which is around 650 mph at the latitudes we flew at. I'd not thought about this too deeply before it actually happened to me for the first time, but the inevitable consequence of us flying faster than the Earth's rotation was that we became witnesses to something few people had ever seen: the sun rising in the west.

Soon after accelerating past the southern Irish coast, we'd hit Mach 1.5, the speed at which we were flying 50 per cent faster than the Earth's rotation speed. After what seemed like endless darkness, I would suddenly see the faintest glow on the horizon. Where, moments earlier, it had been pitch-black, it now grew lighter and lighter – and then, all of a sudden, the sun began to rise. This wasn't a 'conventional' sunrise, however. It was a sunset in reverse, with its own unique hues. Eventually, it would get to be broad daylight again.

At certain times of the year, we'd arrive over Manhattan as the second sun was setting again. We'd land, deplane, get on board a bus and head into town. Crossing the bridge on to Manhattan Island, I'd gaze past the Empire State Building where the sun would be setting again – and no matter how many times I saw it, the thought would always be the same: *Good grief, I've seen two sunrises and two sunsets today. How many people get to do that?*

The big things we were selling were punctuality, speed, comfort and regularity – but, on top of this, and subtly, the fact, too, you'd feel so much better flying across the Atlantic on

Concorde than you would on a subsonic aeroplane. This wasn't something we could tout openly because the Atlantic was a major conventional market for us. But it was true – and we knew it because there was science behind it. Concorde's fuselage could be pressurised to 11.7 pounds per square inch. The cabin altitude this gave you at 58,000 feet was of the order of 5,000 feet; whereas, on a 747 flying at 35,000 feet, the cabin altitude was 8,000 feet. If you were flying Concorde, you were in a machine that was the equivalent of being on top of a hill; the others were on top of a mountain. Furthermore, you're up that hill for three hours and the other guys are on their mountain for more than twice that. This made a big difference to our passengers' well-being. Add to that the fact you're travelling faster than the Earth rotates – rolling the sun backwards – and it did something magical, we discovered, to the body's clock, in effect, resetting it. So, coming in to land at JFK, after the equivalent of being up a hill, it really did feel like you'd just got up and it was 9.30 in the morning. People knew about the punctuality and the speed – they were the selling points. But when passengers started regularly to travel on the aircraft, they appreciated that there were other benefits – the stuff we didn't talk about.

And, of course, the crew got the same benefits. Concorde cabin crew were unusual in that they were part of BA's short-haul fleet. Cabin crew had to have indicated that they were interested in flying on Concorde in order to have been considered in the first place. You also had to be tipped as a particularly good cabin crew member, whatever that meant. And to commit to flying the other short-haul routes, which further delineated those who were selected – not all cabin crew wanted to go from flying long-haul with all the benefits that conferred to short-haul, even if it meant that Concorde was a part of the mix. For most cabin crew, Concorde was generally a minimum two-year commitment, but also a maximum three-year commitment, which ensured that nobody got stale on the job.

The short-haul work pattern for cabin crew was predominantly a rotation. Six days' work and three days off. This was an attraction of working short-haul in that the six-day/three-day pattern was sacrosanct – you could plot months in advance when your 'off' days were, because they were set in stone. Of the six days on, four would probably be based around domestic and European sectors and the last two on Concorde. The 6/3 pattern, and the limited route structure, needed a small group of cabin crew, which meant there was a strong likelihood you'd fly with cabin crew you knew. This led to strong bonds being formed – good friendships – outside of the airline and the aircraft and helped further to mould a strong *esprit de corps* amongst those who flew on the aeroplane; it added, too, to the 'club' atmosphere on board. Many passengers told me that Concorde was just like a club – albeit a very exclusive one. Many more, on stepping aboard, say, at JFK, would say things like, 'I've had a really long, hard six days of work, but when I step on to Concorde, it's like being home.'

The initial deal struck between the company and long-haul cabin crew was that whenever they flew a supersonic sector they had to have a clear day off the following day. That expensive arrangement disappeared when Concorde became a 'short-haul' aircraft – at least as far as cabin crew were concerned.

But we still didn't fly 'there and back' in a day, as some assumed that we did. Flying out of London, the first Concorde took off at about 11:00 and arrived in New York at about 09:20. It then sat on the ground until 13:45, when it took off back to London, landing at Heathrow at around 22:00 hours. The aircraft, then, was away from London for approximately eleven hours. Add briefing and debriefing time and a crew would have been working for thirteen hours a day, which, under the rules was barely half an hour short of a maximum duty day. If the aircraft were to suffer a delay, the last thing we wanted was for the crew to run out of hours – not when you're selling speed and punctuality to your customers. So, if we had planned to go 'there and back' in a day we would have needed to position a standby crew in New York to cover that eventuality and any cost advantage would therefore disappear.

The thinking, then, was that the crew coming off the flight that landed in New York at 09:20 were better utilised staying over, and then flying back to London the next day.

On the very rare occasions when there was a crew member taken ill in New York, cabin crew could operate with one missing and inbound flight crew could be asked to go 'there and back' as a one-off 'in extremis' exercise. I did that a few times in my twenty-two years and it was a strange experience. Get out of bed, have breakfast, go to work, come home, have dinner and back to bed. And I'd been to the USA and back, having popped into my office at Heathrow, picked up the contents of my in-tray and worked on it whilst I'd been on the ground at Kennedy Airport.

This crew rostering pattern meant that we got all the benefits that the passengers got. Because of the aircraft's inherent reliability, and the nature of the service we were offering, I could make New York restaurant reservations and theatre bookings months in advance with great confidence that I'd be able to honour them. Others got to be even more adventurous. One group of crew members owned a car, which in the winter they'd run upstate for the day to go skiing. Another group clubbed together and bought a light plane that they kept at nearby Teterboro Airport – their jumping-off point for all kinds of excursions in the US. Other crew members signed up to golf clubs. Two more bought an apartment. One even lived in New York.

But it was, first and foremost, the passengers who reaped the real rewards of supersonic travel – and, right from the start, I found myself rubbing shoulders with a swathe of the great, the good and, just occasionally, the less-than.

20

I'm a fan of the ballet – always have been. Soon after I started as a co-pilot on Concorde, I found myself in London and, my business for the day finished, wandered up to Covent Garden on spec to see if the Royal Opera House had any tickets available for that evening's performance. I didn't even know what was on, but it happened to be *Hamlet*. The box office told me I was in luck, too, because there'd been two returns moments before I'd shown up. So, grateful for my good fortune, I bought them both – my girlfriend at the time, also being a ballet fan, was rushing toot-sweet to meet me before curtain-up.

At the interval, we were raving about the performance over our glasses of wine – and in particular about that of the male lead.

'He's as good as Nureyev, I reckon,' I enthused to my girlfriend.

A man standing at the bar next to us laughed, nudged me and said: 'I think you ought to know, it *is* Nureyev.' He and his wife were particularly amused (or maybe they were just saying this) because they'd booked their tickets a year beforehand and we'd rocked up on the night not realising it was Nureyev.

Fast forward a couple of years and Nureyev was on the plane and I invited him onto the flight deck. I began relating the story to him, but mid-flow I stopped and asked him what his favourite ballet was. *Hamlet*, he told me. And then, in a moment that somehow felt pre-ordained, he added quite un-*apropos* of anything I'd said: 'My best-ever performance was at Covent Garden.'

I told him then I'd been there that night. He replied with a smile: 'Well, then, we've now both watched each other at work.'

Magical things seemed to happen on that aeroplane. I'm a big golf enthusiast, too, and not long afterwards Tom Watson was on board. We were flying the entire US squad back to New York after the British Open – being a golf fan, it was one of the reasons I'd bid for and got the trip. Knowing we were all golf enthusiasts, Watson brought the Open trophy up on to flight deck. We sat there and had our pictures taken with it and him. Watson was an all-round nice, easy-going guy. He sat up with us talking, until a stewardess appeared to tell him that his meal was about to be served in the cabin.

'OK, guys, I'm going to go back and eat my lunch,' he said, 'but why don't you hang on to the trophy for me until I'm back. Can't think of a safer place for it to be . . .'

Out came everyone's cameras so we could each have a photo of ourselves holding the world-famous Claret Jug. My go first. Thirty seconds after he'd left the flight deck, I dropped it on the centre console, putting a not inconsiderable dent in it. When I saw the trophy again about ten years later, the dent was still there.

The following year, Watson was back again, only this time it hadn't gone so well for him. He'd been in pole position on the seventeenth but had hit the ball beyond the green and ended up with an almost unplayable shot. The mistake cost him dear, and he finished second. Part-way across the Atlantic, one of the stewardesses came on to the flight deck with a message: 'Hi Mike, Tom Watson sends his compliments and wondered if he might come up and see you again?'

'Sure,' I said, 'send him along.'

I was still a co-pilot, in my early mid-thirties, so the fact that the great Tom Watson had asked for me personally made my chest swell a bit. The dented cup incident had long been forgotten (at least by us). Eventually, Tom appeared and sat down in the fourth seat.

'Hey, Mike, how are you?' he said easily.

I could sense the captain and the flight engineer bristle a bit.

'Doing great, thanks, Tom. Good to see you again.'

After a few more pleasantries, he asked if I'd sign his Mach 2 certificate, because in all the excitement last time, he'd forgotten to get my autograph.

The cabin temperature, chilled to keep us and the instruments nice and cool, dropped a notch or two. This wasn't how things normally happened – the whippersnapper getting all the attention. The temperature rose again when Tom eventually asked everyone on the flight deck to scribble their name on the certificate.

We carried on chatting and before long found ourselves approaching Newfoundland. Within a few more minutes, we were talking to Gander ATC on VHF.

Tom nudged me. 'Mind if I listen in?' he mouthed. I gave him the thumbs-up and handed him a headset.

The Canadian controller came back again. 'Hey,' he said, 'are you the flight that's got all the golfers on board?'

'Affirmative,' I replied, turning round to Tom. 'In fact, we've got Tom Watson on the flight deck now, listening in.'

'That's fantastic,' the controller said. 'Can I talk to him?'

I looked at Watson, who nodded enthusiastically.

'Sure,' I said. 'Go ahead.'

'Hi Tom, it's so great to be able to talk to you,' the controller gushed.

Tom said something characteristically demure, and they chatted some more. The controller then said: 'Such a shame about the seventeenth . . .'

'Yeah, well, these things happen,' Tom replied.

My headphones filled with static for a moment, then from across the ether the controller's voice again: 'Tom, can I give you a tip?'

Before anyone could say anything, the controller got right to it: 'What you really need to do is move your hand *around* a bit . . .'

We all collapsed – Tom, great sport that he is, was laughing loudest of all – but the controller was deadly serious.

We, meanwhile, were laughing so much that one of the cabin crew came forward to ask what was so funny because the passengers wanted to know.

There wasn't quite so much laughter when HRH Princess Margaret came up to see us on a flight back from Bahrain.

The Royal Family used Concorde a lot. There were times when they travelled incognito and other times when they used her for state visits; times when they chartered part of the aircraft and times when they chartered the whole plane.

On this particular trip, Captain Hector McMullen and I were due to fly the aircraft out and back to Bahrain on a normal scheduled trip. The unwritten rule says one pilot will fly outbound and the other the return leg to/from the destination. It was agreed that Hector would fly out to Bahrain and I'd fly back. The flight to Bahrain was uneventful, but on the way back, during our briefing at the airport, the briefing officer, BA's in-country manager, tapped Hector on the shoulder and asked if he could have a discreet word. They huddled in the corner for a few minutes. Hector then returned to his seat and nothing more was said.

In the midst of our pre-flight checks on the flight deck, however, Hector turned to me and said: 'I expect you're wondering what that was all about?'

'What?' I said, trying to sound nonchalant as I punched our geographical coordinates into the inertial navigation system.

'Our man in Bahrain and me . . .'

'Oh, that,' I said. 'What about it?'

Hector leaned across the centre console and said, *sotto voce*: 'We're flying Princess Margaret back – her and Lord Snowdon.' Then, before I could say anything, he added: 'They're coming back from an official visit, so I think it only appropriate that I should fly the aircraft back.'

Inside, I fumed quietly, somehow managing to acknowledge with a 'Yes, Sir' through gritted teeth.

Normally, when we invited a Royal onto the flight deck, they sent a polite decline. But, as we were approaching Heathrow, Princess Margaret let the cabin staff know that she would like to sit on the flight deck for the landing.

She duly appeared and sat down on the fourth seat. Hector was still flying, and I was still sulking. As we were landing towards the east, and Hector had his hands full, I thought it might be nice to let Her Royal Highness know that we'd be passing over a number of landmarks she might be familiar with.

The first of these was Ascot Racecourse, which I pointed out to her 6,000 feet below.

'Very interesting,' the princess said, barely moving a muscle.

A few moments later, a few miles to the south-west of

Windsor, I invited her again to look out to the left and down. 'Your Royal Highness, we're now passing over Bracknell at around 4,000 feet.'

This time, the princess glanced left and, looking down her nose, said: 'Whoever wants to know where Bracknell is?'

This might ordinarily have been a bit of a conversation-killer, but I'd been bidden to 'extend all courtesies', so ploughed on. As we swung round on final approach, I informed Her Royal Highness that things were going to get a little busier from here on in, but that we would shortly be passing by Windsor Castle.

A few moments later, as I was in the middle of monitoring our instruments for landing, we got a crackle in our headsets and then: 'Ah! Yes, there's Windsor Castle. I can see my sister's home – the Royal Standard is flying.'

Which, indeed, it was. As the lights of the runway twinkled in front of us, I felt pleased at least that the flight had ended on an upbeat note. But the words had formed in my head too soon. Moments later, Hector slammed us down onto the runway – the main gear hit really heavily, followed seconds later by an equally filling-loosening impact by the nosewheel.

The devil on my left shoulder rubbed his little hands and whispered in my ear that there really was some justice after all.

A deathly hush, meanwhile, descended on the flight deck. Then, all of a sudden, as we were rolling to a halt, the princess's voice chipped in over the intercom for the last time: 'That's his knighthood gone,' she said gleefully.

But of all the people we had on the flight deck at that time, one of them stood clear. He wasn't royalty, he wasn't a celebrity or even one of my sporting heroes. His name was Bill Weaver, and he was the engineering test pilot for the Lockheed L-1011 TriStar, an aircraft BA had had in the fleet for a few years and was now trying to get rid of. Word came down from on high that we had this chap flying with us and that, as was customary for certain special passengers, we were to – to use that magic phrase again – 'extend all courtesies', which we duly did.

Extending these courtesies meant, bare minimum, a visit to the flight deck, but, if we felt like it, to allow them to sit in for the take-off and/or landing.

Bill was flying with us from Washington to Heathrow, so the captain invited him up for the take-off.

We learnt then that BA was seeking to sell its TriStars to the RAF as air-refuelling tankers and that Bill, working for the company that had built the L-1011, was instrumental in helping design the tanker conversion. He was, just as we'd been told, a very nice guy: a tall, easy-going Korean War US Air Force combat veteran, with an open smile and eyes that twinkled when we ran over the technicalities of 'our bird'.

As we were about to settle into our acceleration over the Atlantic, the captain turned to the flight engineer and, in a low, conspiratorial voice, asked if he'd close the flight deck door. I wondered what was afoot. The captain then turned to Bill and, gesturing to the controls, asked if he'd like to 'have a go'.

'I'd love to,' said Bill.

I climbed out of my seat, allowing Bill to climb in. When he was settled, the captain took us out of autopilot, made sure we were stable and then told Bill he 'had control'. *Mmm*, I thought, *I might pay a little more attention to this.*

As Bill took the controls, the captain then began to talk him through going supersonic – which he then went and did. We continued to accelerate, through Mach 1.7 and finally into the cruise. Throughout, Bill flew us deftly and professionally. In fact, something told me we were in very good hands.

Finally, the captain said: 'Right, I'm going to pop the auto-pilot back on, Mike's going to jump back into your seat, and you can go back to the cabin and have a jolly good lunch.' He shook Bill's hand. 'Well done, you've earned it.'

Bill disappeared back into the cabin and doubtless knocked back a couple of glasses of champagne to celebrate his joining the supersonic club.

I asked the captain a number of questions about our charming but mysterious passenger-cum-stand-in-first officer, but if he knew any more, he was playing his cards close to his chest. As we approached London, Bill was once more invited on to the flight deck. I half-expected the captain to invite him to land us, but he didn't – he merely observed from the jump-seat.

At the end of it all, Bill was effusive in his thanks to all three of us for allowing him to fly through the sound barrier. 'That was really good, really good,' he said, 'I just wish I could buy y'all a beer to show my appreciation.'

I looked at our captain, then turned back to Bill. 'Well, it so happens you can,' I told him. 'We often go for a pint at a pub just outside the perimeter called The Crown. We're heading there tonight. Would you like to join us?'

'Love to,' Bill said. He scribbled down the address and let us know he'd be there.

An hour later, Bill wandered into The Crown and, true to his word, bought a couple of rounds of drinks. We ended up sitting next to each other at the bar and I asked if this was the first time he'd been supersonic. 'Because something tells me it's not, Bill . . .'

He smiled. 'You're right. Before I got to do this, I was chief test pilot for a little ol' plane called the Blackbird.'

Several pennies then dropped. The SR-71 Blackbird was the US Air Force's Mach 3+ black reconnaissance jet.

Developed in the 1960s from the super-secret A12 spy-plane of the Central Intelligence Agency, the SR-71 was the fastest jet-powered aircraft in the world – its top speed still a classified secret.

Bill then began to tell me the most incredible story – of the time, while testing the Blackbird, it had disintegrated mid-flight, while he'd been batting along at . . . well, that was the point, he couldn't say; but it had been close to the aircraft's top speed, he said.

'It was a cold January morning back in 1966,' Bill began. We'd been joined by now by several other members of the crew. 'We were flying out of Edwards – Edwards Air Force Base in the high desert. The first part of the test went fine. We'd just topped up from a KC-135 tanker and had accelerated into the cruise.' He smiled again, 'Let's just say we were doing well in excess of Mach 3 and had climbed almost to 80,000 when the right engine started playin' up.'

The SR-71's intake cones forward of its two Pratt and Whitney J58 turbojets functioned a lot like the ramps on Concorde for slowing supersonic air in the intake. The right engine malfunction required Bill to 'go to manual', which he did, but the Blackbird promptly suffered what he referred to as an 'engine unstart' – what we called a 'surge'.

The unstart had sent a shock wave of air – normally positioned at an optimum distance between the lip of the inlet and the face of the engine – forward, which prompted an instantaneous loss of thrust. The first Bill and his colleague – Lockheed flight test reconnaissance and navigation systems specialist Jim Zwayer – knew about it was when a series of very noisy bangs emanated from the right engine and the Blackbird began to yaw violently.

Just prior to this, as part of the test, Bill had begun a thirty-five-degree bank turn to the right. A properly functioning automatic system would 'recapture' the unstart, but this, he said, wasn't happening. The aircraft began to roll to the right and started pitching up.

'I jammed the control stick as far left and forward as it would go, but got no response,' Bill said. He knew then that he was about to be in for what he termed 'a wild ride'.

He attempted to tell Zwayer to stay with the aircraft until he could reduce speed and altitude for an ejection, which at the height and speed they were flying at would almost inevitably be fatal. But, with the rapid onset of g-forces he couldn't even get the words out. And that, pretty much, was the last thing he remembered until coming to in the rarefied air of the stratosphere, the sound of wind rushing past his ears and what sounded like straps flapping . . .

As he told it to me and subsequently,* the aircraft had literally broken apart all around him.

One moment, he and Zwayer had been flying along well in excess of Mach 3, the next he was falling free in almost airless space. It had taken three seconds for the Blackbird to

* My memory of Bill's story that night has been helped by his own words of what happened at the Roadrunners Internationale website: https://roadrunnersinternationale.com/weaver_sr71_bailout.html. Sadly Bill passed away on 28th July 2021, aged ninety-two.

transition from a flight regime in which everything had been normal to it breaking up. The rapid onset of the high g-forces had caused him to black out.

As he came to, Bill's first thought was that he couldn't possibly have survived the break-up of his aircraft – not at that height and speed – therefore he had to be dead; and since he felt euphoric, being dead perhaps wasn't so bad after all.

Fortunately, his suit – much like the pressure suit astronauts wore – had inflated; the emergency oxygen cylinder in the seat kit attached to his parachute harness had done what it was supposed to: not only providing him with oxygen, but maintaining pressurisation, preventing his blood from boiling at the extreme altitude.

But as his awareness returned, he realised his problems were only just beginning.

His high-altitude egress from the aircraft meant the air around him was insufficiently dense to stop his body tumbling, but as he wasn't tumbling, he figured that the drogue chute, designed to stabilise his body in freefall, must have been working. It was impossible to gauge anything beyond the fact that he was alive because his helmet visor had iced up – he had no idea how high he was from the ground; whether he was past the point – 15,000 feet – where his chute was supposed to open automatically.

'I felt for the manual-activation D-ring on my chute harness, but with the suit inflated and my hands numbed by the cold, I couldn't locate it,' he said. 'I decided I'd better open the face plate, try to estimate my height above the ground, then locate that D-ring.' But before he could reach for the face plate, the main chute deployed.

'I saw I was descending through a clear, winter sky with unlimited visibility.' He was also greatly relieved to see Zwayer's parachute coming down about a quarter of a mile away.

Because the SR-71 at that speed and height had a turning radius of a hundred miles, all he knew was that he was somewhere over the New Mexico, Colorado, Texas, Oklahoma border region.

He landed unharmed in the desert scrub. As he struggled to collapse his billowing chute, he was astonished to hear a voice behind him. 'Can I help you?' it asked.

He looked up to see a guy in a cowboy hat walking towards him. Even more astonishingly, there was a helicopter behind him, its rotors gently idling.

The gentleman's name was Albert Mitchell Jr and Bill had come down a mile and a half from his house. Mitchell owned a huge cattle ranch in northeastern New Mexico. Seeing the two chutes come down, he'd radioed the New Mexico Highway Patrol, the Air Force and the nearest hospital – to which, he said, they'd be headed as soon as he flew off to check on 'the other guy'. Ten minutes later, Mitchell returned with the devastating news that Jim Zwayer was dead – his neck broken from the Blackbird's disintegration.

Although deeply saddened by his colleague's death, Bill marvelled at the miracle that had saved him. His ejection seat had never left the aircraft; he'd been ripped out of his seatbelt and shoulder harness by the extreme forces.

After visiting his colleague's body and gaining Mitchell's assurances his ranch foreman would watch over it until the authorities arrived, they both set off for the hospital in Mitchell's little Hughes helicopter. The rancher flew the machine at the very extremity of its limits, Bill explained, terrifying him that it, too, would disintegrate before they ever reached the hospital. But somehow it kept going.

When they touched down and he was transferred into the hands of the hospital staff, Bill paused to thank his Good

Samaritan, asking him as he did so if he'd had many hours on the Hughes. 'Nope,' the hardy Mr Mitchell replied, 'this is the first time I've flown the darn thing since the last crash.'

Bill finished, leaving me open-mouthed. Earlier I'd asked if this was the first time he'd been supersonic. He had, in trumps. Almost twice as fast as I'd ever flown. Mach 3.5+ – and without an aircraft! Sometimes you feel you've asked the dumbest of questions . . .

Outside the pub, I asked Bill what he had attributed his remarkable escape to.

'I was lucky, pure and simple, Mike.'

Then he drew breath, paused for a second and added: 'Respect the airplane and she'll respect you back. That figures in there somewhere, too.'

He smiled. 'But then, I hardly need tell you guys that, huh?'

He waved down a passing cab, clambered in, tipped us a finger and was gone.

On my way home, I thought about what Bill had said.

Beyond the quality of his extraordinary *Right Stuff* tale, there were some salutary lessons in it for all of us.

Because it flew so fast, the Blackbird employed unusual methods for adjusting trim-drag, as did we. On the day of the test, Bill and Zwayer were investigating procedures to reduce trim-drag and improve high-Mach cruise performance. This involved adjusting the Blackbird's centre of gravity further aft than normal.

Like Concorde, the SR-71 relied on a fuel-transfer system to shift its C of G. In moving it aft of its normal limits, the Blackbird's longitudinal stability had been reduced, making it unstable. Whether or not this was responsible for the accident, in part or whole, testing the Blackbird's C of G aft of normal limits was discontinued – from then on, all trim-drag issues were resolved by aerodynamic means.

Two weeks after the accident, Bill was back flying an SR-71 again, epitomising the *sang froid* for which he and his brotherhood of test pilots were known.

Concorde flew in a corner of the flight envelope that was known to a small number of aircraft designers – amongst them the virtuosi at Lockheed who had designed the Blackbird.

If ever a reminder had been served regarding the hostility of that environment, Bill's story was it.

He had been in a military aircraft – he and his flight systems engineer had been clothed in full pressure suits.

We flew in shirtsleeves – and we carried a hundred passengers behind us who were able to sip champagne and eat lunch within what seemed like touching distance of space. But up there, air travel was nuanced. Our aircraft – like the Blackbird – was a high-maintenance thoroughbred.

Respect the airplane and she'll respect you back . . .

What did that mean in relation to the aircraft I was privileged to fly?

Thanks to the miracles of engineering behind all the systems and subsystems that had come together to enable her to *be*, we had an aircraft in our possession that flew – as the Blackbird did – in the boundary layer between Earth and space. That she continued to do so – day in, day out, thus far without incident – had been due to our adherence to Bill's adage: we respected the aircraft utterly; and she'd afforded us that same deep courtesy.

Bill's SR-71 had crashed because it had flown outside its normal limits – chillingly, in a foreshadowing of what would later happen to Concorde, it had become unstable and had broken up when its C of G had been altered from a setting, which, until his fateful flight, had worked perfectly.

As I pulled into the drive of my home, I was reminded of something that had come to me when I'd been presented with the file on Speedbird 911 – the BOAC 707 that had broken up over and crashed on the slopes of Mount Fuji.

Respect the process. The things we did in and out of the flight deck had been built on those all-important 'foundations of understanding'. It was one of the reasons we had checklists. It was one of the reasons, too, that everything that we did in the cockpit, we did after we'd discussed and agreed it.

The system thus far had worked. But to paraphrase Bill, if we ignored or abused the system, the consequences had the capacity to prove fatal.

21

Prior to the fuel crisis of 1973–4, BAC and Sud Aviation had 200-plus orders for Concorde and the future of the aeroplane looked moderately rosy, with many airlines thinking this was the way to go. But when the fuel crisis hit, they all reconsidered. The choice they faced was binary: do we buy this supersonic airliner that can carry 100 passengers but only as far as the width of the Atlantic? Or do we buy this big 747, which has just come out, which can carry up to 450 passengers pretty much anywhere in the world? Ironically, you'd make more money flying a full Concorde across the Atlantic than you would a full 747, but there was another element to consider: your customer base of the future. To the masses who were predicted to fly in the era that followed the fuel crisis, Concorde seemed even more elitist than she'd seemed before.

So, every single airline decided that they didn't want Concorde – including BA and Air France. All orders and options were cancelled. BA and Air France wanted to cancel theirs, too, but the respective governments behind the two airlines pointed out that they, the state, *de facto* the taxpayer, owned the airlines – and presented them with some cold, hard facts. No matter what the airlines said, the two governments were resolved – they wanted them to continue to operate the aircraft.

After some more bleating from the airlines about

profitability, a compromise was reached: we'll underwrite your losses, the two governments said.

There was a further issue to consider.

All aircraft need ongoing technical support from the manufacturer for the duration of their operational lives. This carries a cost – the cost of the infrastructure – the plant, the machinery, the employees and other ancillary overheads – associated with keeping the aircraft in the air. If you wanted the plane, you had to be prepared to pay a premium on the sale that covered this cost. If you sell 1,600 airliners a year – as, before the Covid crisis, Boeing and Airbus had been achieving for many years – the added premium is relatively small. But if the only number of aircraft you were ever going to sell was as low as fourteen – as was the case with Concorde – that premium is pretty horrendous.

It was agreed in the UK that the taxpayer would pay for Concorde's infrastructure costs, but the government, not unreasonably, put a rider on this. If you, BA, make a profit, then the British government, *aka* the taxpayer, will take half of it.

At the time, corporation tax was 50 per cent. This meant that for every pound that BA made, the government would take fifty pence to cover the aircraft's infrastructure costs, then another twenty-five pence in corporation tax. There was, thus, no incentive for BA to make any money on it – and this was true, too, at Air France. In actual fact, though the aircraft was expected to lose money hand over fist, it didn't because of the initial novelty and exclusivity factor, on top of the core element – the aircraft's utility as a business tool. In the early years, therefore, it did lose money, but not as much as the doom merchants had forecast.

When Sir John, later Lord, King came in as chairman of

BA, he was determined to put this state of affairs right. As part of a deal the government of the day cut to take BA out of public ownership and into the private sector – that government being the Conservative new broom of Margaret Thatcher – King gave his management team an ultimatum: two-and-a-half years to turn the aeroplane around and make a real profit, otherwise it'd stop operating it.

In the deal that transpired, BA picked up everything the government had previously been underwriting – the liability for the losses that had previously been taken on the chin by the taxpayer; and all the infrastructure costs, at the time in the order of £18 million per year – a sum that BA needed to pay the manufacturer before any of its Concordes moved an inch. The new numbers suggested a profit could be made on this basis, its flipside being that *if* it were made, BA kept it – no more 50 per cent handed to the government; corporation tax, of course, would continue to be paid in the normal way.

All of a sudden, alleviated by the removal of the 50 per cent, profit margins hoved into view.

At the same time, BA picked up chunks of responsibility for some other overheads: the simulator, for example, which had previously been owned by the manufacturer; the spares inventory, which had a sizeable capital value that impacted favourably on the balance sheet; and the remaining flyable test aeroplane, G-BBDG, the last of the six aircraft to fly in the test programme. Finally, for a nominal amount each, two other aircraft were brought into service – making seven flyable aircraft across the BA fleet (*Delta Golf* not being one of them, but it would be held for spares).

Two people then stepped forward to revamp the Concorde operation so as to make this new situation work.

One of them was a Concorde captain, Brian Walpole; the

other was Jock Lowe, a Concorde first officer. They put themselves forward because they were Concorde enthusiasts –
and, they told John King, they had a plan for making the aircraft profitable. King then gave them their two-and-a-half
years. The clock was ticking.

The first thing they did was to establish a base for the cost
of the tickets. Up until then, the cost had been established
on what was called a 'first class-plus' basis – the 'plus' being
an additional $500 on the cost of a first-class fare. This had
been forged in an agreement with other airlines, mainly in
the USA, when the aircraft came into service – US airlines,
in particular, being frightened that Concorde would steal a
sizeable portion of their very profitable first-class market.

What Brian and Jock did next was to survey their customers to ask what they thought they were paying for a Concorde
ticket. The customers were predominantly chief execs, chief
financial officers and chief operating officers of large companies, who, of course, had never personally paid a penny of
their Concorde tickets. Instead, their companies and PAs
had. As a result, they only had a perception of the value of a
Concorde ticket. When they were asked what that perception was, it came out significantly higher than the actual cost
of a ticket – as much as $2,000 higher (at the time, the cost
of a one-way Concorde ticket was around $3,000 across the
Atlantic). So, what did Brian and Jock do? They implemented
an immediate price hike in the cost of a ticket to the perceived price. They then performed some financial wizardry
with the spares holding to improve the state of the balance
sheet before taking a close look at the operation itself.

What they observed – this being 1984 – was a scheduled
operation comprising two daily flights from London to
New York and three weekly flights between London and

29th November 1962 – The 'Anglo-French Agreement'. The Concorde story starts here. Geoffroy de Courcel signs up for France and Julian Amery for the UK.

The First flight. The UK and France both have their own first flights and flight test crews. France goes first on 2nd March 1969 with André Turcat at the controls.

France's André Turcat and Britain's Brian Trubshaw in full flight test gear at the controls of the Concorde prototype.

Pan Am selling themselves as a prospective Concorde operator. At one stage there were hundreds of orders from airlines around the world. All but BA and Air France were cancelled.

Welcome to tomorrow

An artist's impression of Concorde in BOAC livery. This never happened as BOAC and BEA were folded into British Airways before Concorde came into service. It looks great but the dark blue nose might have led to heating issues.

UK-assembled pre-production Concorde G-AXDN was a key part of the very extensive test programme.

Faster than the sun. Concorde and a solar eclipse.

The Boeing 2707 SST. In 1971 the US abandoned plans for a bigger, faster supersonic aircraft of their own because it was too complicated and too expensive. The prototypes were never completed, but they spent more money on them than the UK had on Concorde – which was already flying.

The Tupolev Tu-144. The Soviet Union also tried to get in on the act. But it was not as powerful, as fast or as capable – despite the PR claims. Their SST looked so much like Concorde that it was nicknamed the 'Concordeski'. Rumour had it that there was industrial espionage going on – but that Concorde's manufacturers deliberately fed them false information.

The two pre-production test Concordes pose together in airline livery. F-WTSB, now at the Aeroscopia Museum, in Toulouse, and G-BBDG, at the Brooklands Museum, in Weybridge.

Me as a boy at Dunstable Grammar School in 1962 – already with a burning ambition to fly.

My first solo flight in a Cessna 150 at Luton Flying Club. It was part of an RAF Special Flying Award that I had been granted.

The 'Squadron' of DHC-1 Chipmunks at the 'College of Air Training' in Hamble. I loved every minute of my 52 hours flying this delightful aircraft.

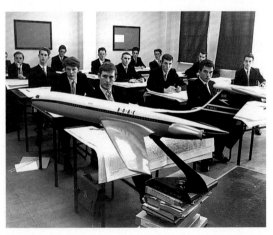

A studious lesson in a Hamble classroom, overseen by a Vickers VC10. It was always my aspiration to fly the all-British aircraft built at Brooklands in Weybridge.

The BOAC VC10. Elegance and speed. So popular with pilots and passengers alike. The VC10 once held the record for the fastest subsonic Atlantic crossing.

A symbol of what it was all about. My first airline wings, awarded when I realised my dream and joined BOAC.

Some of the BA Aircraft flown by me early in my career. I was qualified on the VC10 and 1-11 – and generously 'allowed a go' by captains of the Trident and the Chinook (but don't tell anyone!).

18th October 1982. BA and AF Concordes land simultaneously at Orlando International Airport in Florida to celebrate the openings of the British and French pavilions at the Epcot World Showcase in Walt Disney World.

Her Majesty Queen Elizabeth II aboard Concorde. When she flew, special rear-facing seats were fitted for her *en route* comfort and convenience – plus a walnut desk.

The BA Concorde cabin looks snug when compared to wide bodied-aircraft of today, but was a similar size to its contemporaries – the Boeing 707 and the BAC VC10. However, there was so much more leg room – and you were only seated for just over three hours.

Concorde and a Braniff International Boeing 747 at Dallas Fort Worth International Airport (DFW). Was the speedy Concorde or the massive Jumbo the future? Although Braniff did, for a short time, fly Concorde between Washington Dulles and DFW on an exchange deal with BA and AF, it turned out to be the latter. For now at least . . .

Date	Aircraft			Holder's Operating Capacity	Journey or Nature of Flight				Flying Times					Instrument Flying
	Type	Markings	Captain		From	To		Departure	Arrival	Day In Charge	Day Second	Night In Charge	Night Second	

(handwritten log book entries — first Concorde flight dated 8.8.77)

My handwritten log book showing my first Concorde flight on 8th August 1977, below my last VC10 flight on 16th April that year.

The Concorde cockpit. A cosy home for over a quarter of a century. It's all analogue. In the mid 1990s we did look at fitting modern computer-driven screen displays. But '*if it ain't broke don't fix it!*'

G-BOAB – my first Concorde. It is now situated on the southern side of London's Heathrow Airport on view to millions of passengers each year.

Lift off at around 250 mph with condensation forming in the low air pressure areas generated by the wing's leading-edge vortices.

The sun rising in the west, and shining through Concorde G-BOAE's visor, as we travelled faster than the Earth rotates between London and Boston on 8th October 2003. The flight set a new record for a flight from Heathrow to the USA – just 3 hours 5 minutes and 34 seconds.

Two planes salute each other on the occasion of Concorde's first scheduled service to Washington Dulles on 24th May 1976.

Landing back at Heathrow on runway 27L. Without the drooping nose the view would be obscured. It's a good thing if the pilots can see the runway!

Washington D.C. A service that had existed for a while between London and Singapore, and the service to and from Bahrain, had by now ceased.

In totting up the number of flights between these destinations, Brian and Jock realised they didn't need to be serviced by seven aircraft.

Not unnaturally, BA's Engineering Division, knowing that there were seven planes at the airline's disposal, had geared their plans and operation around them, giving lots of leeway.

The two determined to use up that leeway in a profitable manner. There were some initial objections from Engineering, but the maths, as laid out by Brian and Jock, was inescapable.

One aircraft was required to go from New York to London and back in a day.

A second went to New York in the evening and came back the following day.

A third went to Washington three times a week.

A fourth was required on standby. A fifth, meanwhile, was scheduled to rotate through heavy maintenance.

Which left two 'spare'. Brian and Jock's view was that this was crazy: we had two aircraft that could be making money for us if we could get them up and running.

These two were about to become the difference between an operation that broke even and an operation that was about to become handsomely profitable.

To begin with, under Brian and Jock's oversight, BA began looking at different scheduled operations and new destinations. The first of these was Miami and Dallas/Fort Worth. By transiting Washington and going on to Miami and DFW,

Concorde could still get there quicker than was possible on a conventional flight.

Both these new operations were implemented, the latter in conjunction with US airline Braniff International, which would pick up the flight from D.C. and take it to DFW. The aircraft would remain in Texas overnight, then come back the next day, all of it operated by a Braniff crew. To be able to do this, the aeroplane had to be certificated on the US register, because it was being operated by a US airline between two points within the USA. The US Federal Aviation Administration insisted on some mods to the aircraft, in particular to the way the hydraulic system operated under certain failure conditions.

These, as it turned out, were really beneficial to the overall design.

The other market that Brian and Jock now began to eye up owed its genesis to an unlikely spot on the map: a pub called The Bell Inn at Aldworth, near the Berkshire market town of Newbury. The Bell was the local of a Concorde captain named Brian Calvert, who canvassed the pub's cricket team, its family and friends to see if there might be enough interest amongst them all to charter a Concorde for the day. It turned out there was. Calvert then approached Brian Walpole and Jock Lowe and, after much to-ing and fro-ing, the idea of a Concorde flight for The Bell cricket team and friends was seen as doable. The question for BA then became: what price do we pitch it at?

The itinerary was built on a nominal run around the Scilly Isles, 28 miles off the Cornish coast.

Following take-off from London Heathrow, Concorde would go supersonic, get up to Mach 2, make a long, leisurely turn around the islands whilst champagne and a light lunch was served. It would then head back to London after a round-trip of an hour and forty minutes, landing back at LHR 'in time for tea'.

For this, Brian and Jock calculated a price of £100 a ticket. They put it to Brian Calvert and then sat back and waited to see if there would be any takers.

The flight sold out in a few hours.

After Brian Calvert organised another couple of flights on the same basic principle, it began to dawn on all involved that there was a market here. Further, that we had a spare 'hull' to make it work, because we'd be using an aircraft that would otherwise be sitting around doing nothing. The beauty of this market was that all the fixed costs had already been paid for by the scheduled operation – the charters would be icing on the cake.

All these 'round-the-bay' trips had to account for were the direct operating costs associated with them: the crew, fuel, landing charges, a few other ancillaries.

Several companies grew up around these new operations.

The biggest and best-known was a company called Good-wood Travel, which had started out as a high-end charter operation taking motor enthusiasts from the UK to Monaco. Goodwood's was a visionary operation, because it didn't just think about whizzing passengers 'round the bay' – pretty soon, it was coming up with some imaginative destinations people really wanted to travel to – the value proposition being a simple one: Concorde, if you so wished, could take you there and back in a day.

The kinds of destinations it proposed were Venice, Hel-sinki, Jordan and Cairo. In Venice, you could hook up with the *Orient Express*. In Helsinki, you could see the Northern Lights. In Jordan, you could clamber on to a horse and ride to the lost city of Petra. And just outside Cairo, you could gaze on the Pyramids and the enigmatic Sphinx.

This side of the enterprise became hugely successful. Eventually, it operated to fifty-two destinations and transported around a quarter of a million passengers – all of them on flights chartered predominantly with BA (a few trips were made with Air France).

Lord King's edict that the aircraft had to be profitable within two-and-a-half years had been realised. There were times, in fact, when the seven aircraft in the fleet would contribute around 40 per cent of BA's entire profits – I was told towards the end of my career that it had contributed £500 million net to the airline's bottom line.

That's around a billion in today's money, at a time when that actually meant something; when it bought you more than just two or three footballers.

22

Air France, the French flag carrier, left its arrangements with its government pretty much in place.

Air France's Concorde operation was much smaller than BA's. It didn't have as many scheduled destinations as we did, nor did it have the frequency of flights.

Other differences set us apart, too.

BA pitched its Concorde experience as a 'First Class-Plus' service; Air France advertised its service as 'Business Class-Plus'.

There were rumours, too, that the French government was subsidising Concorde's loss-making operation by funnelling money to Air France from infrastructure support costs supposedly set aside for Paris's new Charles de Gaulle airport – rumours, it should be said, that were never substantiated.

With BA privatised, and the Lord King edict that we needed to make Concorde profitable uppermost in the minds of managers and planners, we adopted a more gloves-off approach to the sibling rivalry than had hitherto been in place between our two airlines and our two Concorde operations.

We identified people living in France who were using Concorde to travel to the USA, but who weren't within driving distance of Paris. These customers had to fly to Paris to connect with Concorde at CDG and, therefore, we surmised a proportion might prefer to fly to London Heathrow to connect with our 'first class-plus' Concorde service instead. We surmised correctly; many of them did.

These things came together during the period I flew Concorde as co-pilot.

Whilst much of what was happening at the strategic level was opaque to me, it was a time when morale amongst crews was high, and we took pride in the way the airline was being transformed. Everyone, owing to their passion for the aircraft, was doing whatever they could to make Concorde a success.

Within BA, the charter flights reinvigorated everyone's enthusiasm for the aircraft – and this was especially true of the 'round-the-bay' trips: the hops we made around the Scilly Isles.

Of the 1 hour 40 minutes' chock-to-chock time, approximately 1 hour 10 was spent in the air. After champagne and a light lunch, there was a visit to the flight deck where passengers could see the gee-whizz stuff that took us to Mach 2 and talk to us about our experience of flying the aeroplane. We put on an extra flight crew member whose sole responsibility was corralling passengers onto the flight deck.

For us, it was wonderful to fly people who had bought – or in many cases been bought – a seat on our aeroplane as a trip-of-a-lifetime experience.

The high-end Goodwood-run flights had a similar effect. These were twenty-three-day round-the-world trips, using the same Concorde for the whole journey, and five- or six-star hotels all the way. There was a crew-change halfway, partly because of our industrial agreements covering crew time away from home, but mostly because BA wanted to give as many crews as possible this unique and extraordinary experience, which took in some of the most spectacular destinations in the world. In total, BA and Air France did twenty of these trips.

In 1984, after seven years on the aircraft, I was asked if I wanted to apply as a training co-pilot, thanks, in part, to the vacancy created by Jock Lowe's move into management.

The selection process included the candidate's technical suitability, his (this was just before the introduction of female pilots at BA; the airline's first female pilot, Lynn Barton, wouldn't fly with the airline until 1987) capacity to pass a number of verbal reasoning and mental agility tests, plus an interview in which we were tested on our technical and inter-personal skills.

I was fortunate to have had a number of experiences beyond those on Concorde, which, in retrospect, helped my selection chances. This included time outside BA (with the airline's blessing) in which I had flown for the UK Army parachute team the 'Red Devils'; helped run an air-charter company that operated a fleet of twin-engine general aviation aircraft and my deep involvement with the Guild of Air Pilots.

Even so, when I got the job, I couldn't have been more thrilled. Before I'd turned thirty-five, I would be teaching others to fly this incredible aeroplane – quite an outcome, it felt, for the seven-year-old who'd looked up and determined that was where he wanted to be. I couldn't help but think back, as well, to key moments in my training – learning to fly the VC10 at Shannon and Concorde at Brize.

Would I come across to my trainees in that way my instructors had – those godlike beings who'd flown Lancasters and Halifaxes during the war? I couldn't begin to see that I would, but the shiny new jets that had been in service when I'd joined BOAC were beginning to give way to a new generation of airliners – not just Concorde (which by now had been in service for almost a decade), but aircraft like the Airbus A320: a short-haul twinjet that, thanks to its next-gen turbofans and fly-by-wire flight controls – in which computers took the pilot's inputs and translated them into digital commands that allowed the aircraft to perform at optimum efficiency – looked set to revolutionise air travel in the way that the 707 and VC10 had when they entered service. The A320 entered service in 1988.

The period that followed my appointment as a Concorde instructor was, in many ways, the best time I ever had on her. Later, I would enjoy being in management, but that came with its own responsibilities. The role of training co-pilot came with very little downside. What could be more satisfying than teaching someone who, like me, had seen flying Concorde as the culmination of a deep, long-held dream?

As a training co-pilot, I was involved in four months of the six-month conversion course, instructing on both the two-month simulator module and the two-month base flying operation – the flights I had undertaken at Brize.

Most pilots experienced a dip at some point in their conversion flying – a moment in which you hit a trough from which, it seemed, you couldn't dig your way out.

To be able to help someone through that period and on out into clear skies, literally and metaphorically, was one of the joys of the job – and allowed me to re-engage in my co-pilot role with a passion and commitment that underpinned my experience of the assignment for the next five years.

By 1989, I'd arrived again at a moment when I needed to think about taking a command, an opportunity that would only be available to me on another aeroplane type.

I'd also decided that I wanted to become involved in the management side of things.

This was a difficult decision in one respect because I was still loving the whole Concorde experience. But it was easy on another level because I knew that one day it would all have to come to an end; and I'd had my go and had been extremely fortunate to have been on the aeroplane for thirteen great years – great for me personally, but a time when the fortunes of the aircraft itself had fluctuated.

So, I thought, move on, and don't come back because nobody ever does – not really.

So, I bid for an aircraft command at the same time as putting myself forward for management positions and, pretty soon, the stars began to align for a command on the Boeing 757/767 fleet (the 757 and 767 were unusual in that, though serving different markets, they effectively had identical cockpits, allowing aircrew cross-qualification on them). But before long, I was called into a meeting with the chief pilot of the TriStar/Concorde medium-haul fleet, Roger Price, who dropped some heavy hints that I should hang fire and wait before accepting a Boeing command. 'Just wait and see what happens,' he told me.

Within a few short weeks, I was invited to be assessed for a management position; and shortly after that, was offered the job of Assistant Flight Training Manager on the BAC 1-11 fleet, the 1-11 being the workhorse of BA's short-haul operations at the time. On the one hand, I was delighted – this was an important step-up into the world of management. On the other, I was more than a little rueful – thirteen years earlier I'd been told that I should avoid taking the Concorde co-pilot job because I was within three years of getting a command. Three years . . . time took on a different quality in this job.

But the 1-11 turned out to be a wonderful experience; I enjoyed flying it enormously. And after three years of the new assignment, in which I flew the aircraft as well as running the fleet for BA, I came to appreciate something else: from the time I'd joined BA (or BOAC as it then was) to this moment – i.e. from 1969 to 1992 – I had only ever flown British-built aircraft.

By the early 1990s, the British aircraft industry was a shadow of what it had once been.

Pockets of innovation still existed within the commercial side of the UK aerospace industry, but, with one exception – the four-engine, 70–130-seater BAe 146 (rebranded in 1992 as the Avro RJ series) – the sector had ceded the actual business of building airliners to Boeing and Airbus, the European consortium. The UK still had much to be proud about in its capacity as a builder of systems and engines – Rolls-Royce being one of the 'big three' global aerospace engine producers. But the era that had thundered to the sound of De Havilland Comets and, latterly, Vickers VC10s and Tridents, was all but over. The future clearly belonged to Airbus and Boeing.

This notion bound me even more tightly to the domestic workhorse I was now responsible for.

For a pilot with nigh-on a quarter of a century's flying experience never even to have touched a Boeing had to be some kind of record, I thought.

After overseeing the training operation on the 1-11, I became the Flight Manager for the 1-11 and the A320 for a while – a 'people' job rather than training or technical. After that, I was offered a 'central' job, Manager Business Development. Then on to become the head of communications for Flight Operations – a role that gave me responsibility across the whole BA fleet.

Next, I was slated to move to the post of Flight Manager, Technical, on the 757 and 767. I was also flying the 757/767 at that stage – at last, my first experience at the controls of a jet built by Boeing.

Unbeknownst to me, all of these moves had a grand design behind them – Jock Lowe, who was by now the Director of Flight Operations for the whole of BA, had got me in mind for a particular job and was keen, therefore, for me to tick certain boxes in managerial positions before he was able to offer it to me.

After a couple of years of shuffling me around the board, the grand chess master revealed his stratagem. I was sitting in my office when I received a call from him.

'Mike?' Jock said, getting straight to the point, 'I'd like to offer you the position of Chief Concorde Pilot.'

I was so taken aback that I said, 'Would you give me a moment or two to think about it?' and put down the phone.

Normally, I'd always discussed my career moves with Chris.

Family life is a team game and big decisions should be made together.

Amy was young and about to start nursery. My two sons, James and Robbie, were at secondary school. They lived with their mum, but we got together regularly and they were – and always would be – a very important part of my life. And Chris was also flying – as a Cabin Services Director on BA's 'Worldwide' fleet.

But, in my mind, this seven-year-old boy was now being offered his job of a lifetime. Literally. It was all that I'd ever wanted in my career. The Chief Pilot on BA's Concorde fleet. What would my mum have said if she'd still been alive?

Dad and she were proud, Staffordshire, working-class

folk. When I was a nipper, and Dad had been a cooper in Burton-upon-Trent, Mum had packed goods in a Co-Op warehouse. They had struggled to support me through Hamble. They had known how deep my passion to be a pilot ran.

Should I take the job? I could call Dad. I could call Chris. I could ask my children. But, in my heart, I knew what they would say. So, a moment or two later, I rang Jock back, thanked him profusely and said I'd be honoured.

Truth be told, the biggest decision of my career had taken a nanosecond.

To begin with, I moved across as the Concorde Flight Training Manager. Not long afterwards, the fleet Technical Manager, a captain called Dave Rowland, was appointed to be the Concorde fleet's Commercial General Manager. As a result, I took on his job as well – the training and technical roles being wrapped under a new catch-all title: Flight Manager, Training and Technical, Concorde fleet. That was quickly shortened, maybe because the name badge was too long, to Flight Manager, Concorde.

This very quickly morphed into the role that Jock had envisioned from the get-go: Chief Concorde Pilot.

Unlike most other fleets, with Concorde, if you were on the management team, you were, by default, also involved in the whole commercial scene, because, by that stage, the aircraft was all about the bottom line.

I'd known Dave Rowland for years, and the partnership worked. I like to think that we both oversaw a period of great stability for the aeroplane.

One of the first issues that confronted me was technical but had implications for the longevity of the fleet.

Twenty-five years on from when Concorde had first flown,

bits of the aircraft were beginning to show their age. In particular, the circuit boards in the computers governing operation of the engine air intake ramps – the devices that ensured the right airflow entered the engines during supersonic flight – needed to be replaced.

For us, the issue was about effecting the upgrade – a major technical operation – without triggering a re-certification of the aircraft. A challenge because most of the kit just wasn't made any more.

The objective, therefore, was to implement what was known as a 'fit, form and function replacement', in which 'fit' meant ensuring the replacement modules fitted into the slots and racks already in existence; 'form' that the replacement had, in effect, to be the same bit of kit, only with the benefits that came with a quarter century's evolution in computer chip design; and 'function' that it had to be able to do exactly the same job as had been done before: managing air entering the intakes at the speed of a bullet.

All of which we were able to do.

The importance of keeping the aircraft upgraded to the standards that befitted her still revolutionary appearance didn't just extend to mods of the fit, form and function variety, however.

In 1995, just as I took on the chief pilot role, we implemented a mod that had come about due to some previous incidents involving burst tyres.

Concorde had long been prone to this particular vulnerability. In the early 1980s, the US National Transportation Safety Board (NTSB) – the equivalent of the UK's Air Accident Investigations Branch (AAIB) – had investigated a series of incidents between 1979 and 1981 involving Concorde's tyres – four separate incidents at Air France and one at BA. Following the first blowout at Washington Dulles, the NTSB recorded its serious concerns about Concorde in a memo that highlighted the damage that had been caused to the aircraft: debris from the tyre had struck the Number 2 engine (the left-hand inboard Olympus), punctured three fuel tanks and severed hydraulic lines and electrical wiring. It had also blown a small hole through the top of the wing.

Following up on the NTSB report, France's air accident agency, the BEA, had recommended the installation of sensors that would alert the cockpit crew to a tyre that had lost pressure. The report also recommended beefing up the tyres and the wheel should a tyre deflate during take-off or landing.

Extraordinarily, no action was taken to implement these changes – beyond a general instruction to aircrew to carry out a thorough inspection of the tyres before each take-off.

That was until 1993, when the UK AAIB issued its findings on two tyre incidents involving BA Concordes that year: the first involving G-BOAF – *Alpha Foxtrot* – during a landing at Heathrow in July; the second, three months later, in October.

In the landing incident, the pilot braked after successfully applying reverse-thrust, the crew felt a bang and got a warning light to do with the braking system. After bringing *Alpha Fox* (as we often abbreviated her to) successfully to the gate, some damage to the Number 3 (aka the inboard starboard) Engine and the Number 8 Fuel Tank just above it was discovered. Fortunately, the tank was empty.

The second incident took place on a Concorde preparing to take off.

Upon applying the brakes, the captain sensed the aircraft moving in an unusual way. When he re-applied the brakes, the crew heard a loud bang and multiple wheel warning lights came on, prompting the captain to emergency-brake. With the aircraft at a stop, he opened his side window and looked back to see fuel streaming from the left side of the aircraft. He called for immediate shutdown of the engines. Subsequent inspection revealed that a large part of a water-deflector from the nosewheel had broken away and that this had punctured one of the fuel tanks on the left side of the aircraft.

Afterwards, BA implemented a change to the nosewheel bogie. To prevent its water-deflector breaking away in future, a cable was added attaching it to the structure of the nosewheel.

Bizarrely, however, no action was taken by our French counterpart.

Soon after the crash at Gonesse, in response to media digging into the causes, Air France was forced to explain its reasoning for this omission.

Under French civil aviation rules, it had not been obliged to introduce the mod, it said, adding, in its opinion, it wouldn't have prevented the deflectors coming off.

I wish I could report Air France's oversight had triggered alarm bells at BA at the time, but our primary, if not exclusive, focus was on *our* fleet, not theirs.

Had I been aware of the French decision not to implement the mods, I would have been reminded of the lessons I'd learnt in my early days flying the VC10: the things you do on or off the flight deck aren't always based on the rulebook, but on something else – the piece of wisdom I'd acquired from those experiences with ex-bomber-pilot instructors and my first VC10 captains about those 'foundations of understanding' that led to best practice; the perils that awaited if you failed to stick to them. I would have then, perhaps, sensed a tick-tick-ticking of a disaster in the offing; one that was beginning now to play out.

I wish I could also report that, like those intuitives who'd been gnawed by premonitions before the sailing of the *Titanic* that something terrible was going to happen, I, too, had sensed something was going to happen to Concorde. But I never did.

What I saw instead when I surveyed the hinterland of my new job, was a world in which everything looked good. Like our campaign to broaden the charter market, for example, which had been kickstarted recently by Dave Rowland.

Goodwood Travel, the company that had led the effort to introduce an upmarket charter business, had started a new charter operation in which passengers were flown on Concorde to New York, after some time in the 'Big Apple', where they would jump on the *QE2*, flagship of the Cunard fleet, and sail back to the UK. Americans were offered the same service, only the other way round.

We'd noticed that this charter was becoming ever more popular and with profits in the sector already high, we asked ourselves why we needed a middleman. As things stood, Goodwood and the other couple of charter operators offering this service had to undertake to fill an entire Concorde – 100 passengers. If we set something up with Cunard directly, we realised we'd be able to be infinitely more streamlined, slick and efficient. All we had to do was set aside a number of seats on the aircraft – on any scheduled flight that suited us – to charter passengers. It could be as few as ten or as many as thirty – it didn't much matter. The point was, we'd be in charge and the profits would be ours. It would need some management because the price you paid one-way for a charter ticket would be considerably less than the price some chief exec paid for showing up on the day. But the overall cost of the package – being at the luxury end of the market – ensured that it wouldn't be abused; the initial concern being that business travellers, wising up to this, would book themselves on to a charter flight and then ditch the *QE2* ride on the way back.

To promote this service, BA and Cunard organised what they called the 'Three Captains Roadshow' around the USA. The three captains – one from the *QE2*, another from Cunard's new ship the *Royal Viking Sun*, the third being a Concorde captain – would hit a particular catchment area for the kinds of customers we wanted to attract and then sell the hell out of the proposition. It was, in effect, a sophisticated sales pitch first and foremost to travel agents – and was so polished that on some days the team would present in three locations. Neither Dave Rowland nor I could be spared for over seven weeks, so we split the assignment. For me, it was a whole new experience being placed centre-stage – literally – in the commercial wing of the Concorde operation. Previously, my involvement had been in an adjunct role to my day-job in technical and training. But seeing how the operation worked on the customer-facing side of the business was extremely interesting. It was great fun, too.

As part of the deal, a number of Concorde captains were invited to lecture on board Cunard ships on standard Cunard sailings – be it a *QE2* voyage across the Atlantic or a cruise between Florida and Mexico with Caribbean islands *en route*.

On my first voyage, a cruise between Fort Lauderdale and Acapulco via the Panama Canal on board a ship called the *Dynasty*, I was due to make a number of presentations. These weren't the same as the sales pitches we'd done during the

'Three Captains' tour – they were more about the Concorde narrative, filled with gee-whizz stats and photogenic shots of the aircraft. One variation on this theme talked about how the Beast worked; another about how an aircraft that was often perceived to be loss-making now wasn't.

The trick, I'd been told, was to get a lecture slot as late as possible in the sailing. The alternative meant your anonymity was blown early and you became public property for the entire trip. Given what happened to me on this particular voyage, that wouldn't have been a problem.

I was due to make my first presentation on the trip after a very popular author called Mary Higgins Clark. I turned up shortly before I was due to present and was gratified to see she had filled the auditorium. They were rapt at what she had to say – and all she was doing was talking to them. I had slides! And props. Show and tell! This was going to be a total bloody breeze, I told myself.

It wasn't.

As the applause died down, I stepped on to the stage, feeling good in my captain's uniform. Walking to the podium, the cruise director announced my name and the title of my talk, then, remembering something, added: 'By the way, Mary Higgins Clark will be signing copies of her new book outside the theatre hall.'

It took no more than a minute for the auditorium to all but clear.

Somehow, I stumbled on, due in large part to the support I received from a loyal stalwart egging me on from the front row – my wife, Chris, plus another two couples we'd met who'd become friends. Between them, they laughed and clapped in all the right places, and it seemed to be working because, bit by bit, the auditorium began to fill again. I imagined that news of this wonderful lecture must be getting around the ship.

And then I remembered: next up after me . . . bingo.

I looked down at the front row and saw that my wife and friends had made conical (and comical) Concorde noses out of their programmes and were holding them to their faces.

As I looked at them, they drooped their noses to the floor.

Point made, I said to myself. Time to head for the bar . . .

23

For all those who loved Concorde, there had always been a cadre within BA who'd had it in for her.

The Concorde counter-campaign was subtle and insidious, its precise causes hard to pin down, but at least a part of it stemmed from the perception that the aircraft was loss-making and that there had been a 'sliding-doors' moment – the moment Lord King stepped in with his profitability edict – when the aircraft and BA could have, and should have, parted company. Extending this twisted logic, even though BA had had a successful run since then, any woes the airline had experienced during this period could, and should, be blamed on Concorde.

It was an undeniable fact that the aeroplane would never recover all the research and development money that had been invested in it. In the 1960s, this amounted to approximately £1.2 billion – split equally between the UK and French governments. Some of this money *had been* recouped – but there was no way all of it would be. A little like the US space programme, you could always justify the investment by saying that its spin-out effect into the wider economy was considerable – and this was true. Concorde employed 50,000 people in the UK over twenty years and it migrated technologies into Airbus – in which the UK and its aerospace industry was a risk-sharing partner – that wouldn't otherwise have existed for many more years. These included fly-by-wire flight controls, carbon-fibre brakes, computerised engine controls and many other innovations that gave Europe's fledgling plane-maker a step-up on to the podium where it would compete with the USA's commercial aircraft giants: Boeing and, at the time, McDonnell Douglas and Lockheed, too. And all this in a period – the 1960s and '70s – when the UK aviation sector was in the doldrums of its post-war glory heyday. But everyone knew, too, that the Concorde agreement had been about more than aviation – that it was also part of the UK's price-of-entry into the European Economic Community – forerunner of the European Union – and this rubbed off on the aircraft as well.

There was, thus, a political taint to Concorde that compounded bias against her from the anti-brigade within BA – something we dubbed the 'Marmite factor', after the UK's famously divisive beef extract spread.

Never mind that, through Concorde, BA was well on the way by the end of the 1990s to generating the £500 million in profits that would very largely offset the financial investment that the UK had originally made in it.

Those within the airline who didn't like the taste of Marmite had other reasons to be against the aircraft.

One of these was the old trope that Concorde was a carrier-vehicle for the rich and famous – an airliner for the elite – and that, they said, wasn't BA's core market; far from it.

Yes, they said, we wanted to attract premium passengers, but premium passengers for BA meant people who were prepared to pay for a better standard of product – something BA had put a lot of time, effort and money into under the stewardship of Lord King and his chief executive, Colin Marshall; it didn't mean, nor did it need, a platform that so obviously separated the sheep from the goats. On BA's first-class service, nobody was required to be rich, famous or part of an elite, they said; they just needed to pay the money. And if they wanted to show up wearing jeans and a torn T-shirt, that was fine, too.

Many of us, of course, had put a great deal of effort into making sure that Concorde was more widely accessible. Whilst the 'round-the-bay' flights were profit-makers, they had ensured, too, that Concorde became within the reach of people who had always wanted a trip-of-a-lifetime experience.

But, nonetheless, the 'elitist' tag persisted and by the turn of the century it had come to be associated by the 'Marmite-haters' not merely with the aircraft and its passengers, but with Concorde's crews, all its engineers, and all the people involved in the commercial side of Concorde's business.

This was made worse by the issue of cost apportionment within BA. Each division within the airline had, at the very least, to make its operations break even and preferably make a profit. There was no doubt that Concorde was extraordinarily expensive to maintain relative to a conventional aircraft. We were pumping around six to eight times more engineering input per flight hour on Concorde than we were on a 747. This was largely down to the fact that we had to have a cadre of engineers that was totally dedicated to the aircraft – people who did nothing else. The Boeing fleet, on the other hand, was designed so that an engineer on one Boeing type could often cross-qualify on another, with the result that the engineering input per flight hour on a 747, 767 or 757 was considerably less than it was on Concorde. Anywhere you flew, you could find support for a 747. If you got an engine failure on a 747 in Jakarta, you could get a replacement engine the next day there (because you could 'borrow' one off another airline and pay it back later).

You couldn't, of course, do that with Concorde. Consequently, on the books, if you were going by how she stacked up financially solely inside the Engineering Division, she *was* a loss-maker.

On the other hand, there were any number of people who, like me, were excited by, and emotionally committed, to the aeroplane. She was a source of national pride – for the Marmite-lovers in the airline it was a case of: *Isn't it amazing we Brits have done this and that we, BA, have the privilege of flying her. She sets us apart.*

For these people, the proponents, Concorde wasn't a loss-maker. She made money for BA – a lot of money. Nor was she just for the rich, the famous and the elite – in our range of product offerings, we had engineered a price-point that introduced her to a swathe of the British population; as evinced by any number of 80-year-olds, even some 90-year-olds, who'd pumped my hand on the flight deck, telling me, with tears in their eyes, that their kids and/or their grandchildren had saved up to buy them a ticket that had allowed them to experience this extraordinary aeroplane – to fulfill a long-held dream.

Yes, we understood that it was difficult to support, and that it was hard for Engineering to work the sums in her favour. But, all told, she was a profit-maker for the company as a whole: for every pound that Engineering spent on her, BA was getting £1.50 back.

But, we added, there was one thing no amount of money could buy us with Concorde on our side – something that we had dubbed the 'halo effect'.

The halo effect was what set BA and Air France – but to a larger extent BA – apart from other airlines. There were only two airlines in the world that operated Concorde – ourselves and the French flag-carrier. It made us, well, by definition, not unique, exactly, but as near as dammit. This, we knew, was something that had incalculable brand and marketing value – a factor that was emotionally true, but sadly, also, *factually* true. Because it was literally incalculable, the number-crunchers and bean-counters couldn't put a figure on it – they couldn't assign a value to the halo effect in the books. So, they didn't.

As a result, when Concorde *did* come up for retirement, this was one of the things that worked against it.

If someone had assigned a pound or dollar value to the halo effect, the aircraft would have, no doubt, kept going. But they just didn't. And, sadly, without its totem, its jewel in the crown, there is no doubt that, whatever the actual reasons, BA's fortunes waned without her – but I am getting ahead of myself.

As BA embraced the new millennium, these two camps were a real thing.

There were those who thought that Concorde was good for the company and those who thought she was a burden. The numbers were fairly evenly balanced, and they were pretty static – nobody moved from one camp to another.

The issue was so Marmite that a 'don't confuse me with the facts' attitude seemed to prevail inside whatever camp you happened to be in.

What it all added up to was a passive threat to the aircraft. Concorde had always been subject to these undercurrents, but at the dawning of a new age they were more palpable than they had been for a while. It wouldn't take much, I felt, for the antipathy in the 'anti-camp' to erupt in a more overt way.

None of us, of course, had the least idea that it would take a crash.

On the day of the accident, 25th July 2000, as Chris, Amy and I were in the taxi on the M3 *en route* from our home near Heathrow to Southampton Docks for our sailing on board the *QE2*, I happened to glance left as we were approaching the service station at Fleet, halfway into our journey.

One of the ironies of that day, looking back, was that my holiday was of the busman's kind: I would be singing for my supper on the *QE2* as part of my PR obligation to deliver talks on board.

As we sailed to New York I would make a couple of presentations – what Concorde was like to fly, with the usual gee-whizz stats and a couple of anecdotes about some of the more interesting people who'd flown with us who'd come up on to the flight deck.

Glancing left on the M3, I noticed a large, brilliant white aircraft flying very low, doing what looked like impossibly unlikely manoeuvres for a large passenger aircraft – including a very low level, very steep, full-power banking climbing turn.

It took me a moment or two to realise that what I was looking at was part of the Farnborough Air Show, the trade event that took place every two years in Hampshire's normally sedate skies.

The aeroplane I was watching was one of Airbus's latest, a four-engine A340. But instead of wonder, what hit me was a beat of irrational fear. *What if that magnificent aircraft were to crash?*

Inside an hour, I was heading up the gangplank of the *QE2*, and our bags were already in our cabin, when my pager and phone went off at the same moment. *Strange*, I thought – *I'm on holiday*. I lifted it out of my pocket. 'Call BA. Most urgent.' I rang the dedicated number in the message and it went straight into our Crisis Centre at Heathrow. I'm not even sure to this day who answered it. 'Mike. There's been a Concorde crash . . .'

My world stopped. Who? When? Where? How? My friends? Before I could say anything, the duty manager told me that Air France Flight 4590 had crashed at Gonesse.

The rest of the day was a blur: the call to my traumatised friend, Edgar Chillaud of Air France; the what-do-we-do calls on the way back; the terrible images on the screens in the Crisis Centre of the aircraft on fire; the midnight oil we'd burnt trying to piece together the events; and finally, the next day, my homecoming to Chris and Amy – and my six-year-old's summation of the task that now lay before us: ensuring that this truly terrible occurrence never happened again.

24

Air France had halted its Concorde operations; ours had resumed after our decision to pay our respects to the 113 dead with an overnight cessation.

Even on the first night of the disaster, we had a pretty good idea of what had happened: the flight running late, the debris on the runway that had blown a tyre, the fragments that had caused a fuel leak in the underside of the port wing, and the resultant fire; plus the fact that a piece of aircraft debris, found adjacent to the runway, was being paraded by the French authorities as 'Exhibit A' in the investigation now under way.

Whilst not all our data was official on that first night, we felt we had in excess of 90 per cent of the facts by the time dawn broke the following day – and this was later shown to be the case (in fact, it was shown we had 97 per cent of the facts by the end of the night). Indeed, it was our confidence in these data points that had given us – and the crew involved – the assurances needed for our Concorde operations to resume the next day.

But now we needed to get down into the detail – and already that looked like being an issue.

The problem was that the lead French agency wasn't an air accident investigation authority, as would have been the case had the accident happened in the UK, or pretty much anywhere else. Their lead agency was the *préfecture* of Gonesse, the legal authority in the suburb of Paris where the aircraft had crashed. In any air accident investigation, the lead agency is normally the air accident investigation organisation from the country in which the event has taken place, but it will be supported by a host of relevant others: representatives from the aircraft's manufacturers, from the engine manufacturer and the accident investigation authorities of the countries that had been touched by the disaster; in this case, the UK and France. The whole point of it was to be collaborative – everyone involved united in what I had now come to think of as 'Amy's wish': that nothing like this should happen again.

The two accident investigation authorities were France's *Bureau d'Enquêtes et d'Analyses* and the UK's Air Accidents Investigation Branch – with the AAIB providing support to the BEA as the lead investigative authority.

But, unlike investigations in most other countries, the BEA and the AAIB were subordinate to the Gonesse *préfet* (prefect) and his team, which took some getting used to. The UK equivalent would be a major air accident investigation being run by a magistrate from Hounslow or Kingston had the accident happened the other way around – i.e. had a BA Concorde crashed at or near Heathrow.

This was a criminal investigation as well as an air accident investigation, we were told, and the criminal investigation took priority. The prefect's role was to find out whether there had been any criminal activity and, if so, who should be prosecuted. And this took precedence over all else.

The first thing that the prefect and his team wanted to do was to get the air accident investigators from the UK to sign a legally binding document that said, in essence, that anything we uncovered that was 'of interest' had to be reported to him first. The UK said that that was unacceptable – that it flew in the face of what everyone here believed needed to be done: find the cause and publish it in a comprehensive report that would form the basis of remedial action – ensuring it *didn't* happen again. As a result, for ten days, while lawyers on both sides of the Channel thrashed this out, nothing happened. In the end, we compromised: the UK would report its findings in the normal manner up through the AAIB, whilst in parallel reporting to the prefect – with the prefect having the right to first call on the evidence.

Meanwhile, on the day after the crash, I received a phone call from Robin Tydeman, who was the principal investigator from the AAIB in the investigation. Robin asked whether I would be prepared to act as a chief technical adviser, under secondment to the AAIB. I didn't have to think about it at all – I said yes.

Robin, who was a pilot, asked if BA would be able to get him up to speed on Concorde flight operations – whether he might be able to fly the simulator to prepare him for an understanding of the intricacies of the crash.

I explained to him that the Concorde simulator was at Filton, but that wouldn't be a problem – we could whip down to Bristol the following morning.

There was a momentary pause. 'Actually, Mike, I had another favour to ask. I'm booked on a late afternoon flight to Paris – my opposite number at the BEA wants to take me to the crash site as soon as possible. Your input would be invaluable. I wondered if you'd accompany me. A big ask, but . . .'

This time, my 'yes', though unequivocal, was tempered by a large knot of anxiety. 'Of course. I'll be there. Just tell me where I need to be and when.'

Thirty-six hours after the aeroplane had impacted the ground, destroying most of a hotel, the Hotelissimo, in the process, I stood by the barrier tape staring at what was left of it. Almost two days after the crash, parts of the wreckage were still smouldering. The bodies of the 113 victims – 109 passengers and 4 people at the hotel – had been removed, but it was to them that my mind turned. I had never been to a crash site before, and accompanying Robin on the way, I had not been sure how I would react.

The sight of it hit me hard as I struggled to take in what I was seeing, even when it was explained to us by Robin's counterpart from the BEA.

After taking off from CDG's Runway 26R, Flight 4590 had begun to carry out a left-hand turn, the start of a trajectory that, uninterrupted, would have taken the aircraft towards the centre of Paris, some twenty kilometres away.

Instead, it had crashed just over four kilometres from the end of the runway at a point where two roads, the N17 motorway and the local D902 converged on the edge of the suburb of Gonesse, the point occupied by the hotel.

A whole wing of the building had been destroyed. There was very little to see. I had been told that Flight 4590 had been trying to make for the nearby airfield at Le Bourget and a sizeable part of the tragedy was seeing just how close the crew – Captain Marty, his first officer Jean Marcot and the flight engineer, Gilles Jardinaud – had come to reaching it. To remind me of this, a small executive jet was coming in to land at Le Bourget, whose runway threshold was less than a mile to the south-west. Behind me, on the other side of the road, lay open fields.

Descending ever lower, a large industrial park ahead and to the left, the Le Bourget field within tantalising reach, the aircraft had fallen out of the sky, slamming into the back of the hotel, the fuel in its tanks exploding on impact.

The hotel was all but eviscerated, though a part of the ground floor, where, miraculously, some of the guests had escaped, was still standing.

Of the aircraft itself, initially there was almost nothing recognisable.

Robin nudged me and we dipped under the police tape. A *gendarme* on the other side walked towards us holding his hand up, but Robin showed him his pass and, after satisfying himself we were who we said we were, the policeman waved us through. We strode purposefully towards the part of the hotel that was still standing. Upon reaching it, we climbed the fire-escape stairs. We then surveyed the site.

Now, I could make some sense of the debris field.

Off to our left, I could see part of the rudder, a bit of wing and an engine bay. To the right was the largest collection of bits and components, what looked to me to be part of the forward cabin area. The axis of wreckage – such as it was – ran between these two points and was relatively compact, suggesting the aircraft had been flying almost level with very little forward speed when it hit the ground.

Impact marks next to what was left of a belt of trees, reduced to ash by the fireball, were also visible. But it was impossible to make out anything close by that resembled an aeroplane, let alone what had once been a Concorde.

It was only when Robin drew my attention to an area of wreckage adjacent to the smoking shell of a transformer building approximately fifty yards away that I was able to discern actual piece-parts: the tip of the nose cone, a section of the windshield, a large piece of the nose gear, a part of the rear cockpit and a bit of a seat. A pilot's seat. The section of the aeroplane that had been my workspace – the 'office' I had been so proud to introduce passengers to for all the years I'd flown her – was little more than a smoking assemblage of tiny, molten bits.

After several more minutes, with the smell of burnt fuel, vegetation and molten asphalt rising up from the ground and threatening to choke us, I gazed at the sky and thought again of the crew, the passengers and the victims on the ground. I may even have bowed my head in a moment of silent prayer.

When our BEA host announced it was time to go, it couldn't have come soon enough.

25

An air crash investigation follows set protocols laid down in an agreement called Annex 13 of the Convention of Civil Aviation, also known as the Chicago Convention. This lays out exactly how member states of the International Civil Aviation Organization, a UN-affiliated agency, are supposed to gather and analyse the evidence of a crash, with a specific focus on causes and subsequent safety recommendations.

Within thirty days, the investigators must release a preliminary report to ICAO outlining what they have gathered and concluded, with a final report due as soon as possible thereafter – ideally within a year. If that is impossible, an interim report is expected, or a series of interim reports, detailing progress on each anniversary of the crash or incident, until the full facts are established.

Any investigation is a huge undertaking – a complex jigsaw of pieces that comprise wreckage, analysis of the crash site, witness statements, and a host of background factors, such as crew experience and training and the aircraft's maintenance records.

Evidence from the two 'black boxes' (which are actually orange) are also key.

The flight data recorder (FDR) provides data on the aircraft's critical systems up to the moment of impact. The cockpit voice recorder captures discussions amongst the crew, flight deck sounds and noises, plus air traffic control chat.

Pathologists will examine the crew's remains to establish

whether drugs or alcohol had played any part; and, on the technical/mechanical side, specialists will sift systems, sub-systems and structures down, if needed, to their atomic metallurgical composition for clues – any anomaly – that may point to a possible cause, the overarching aim being improved safety rather than apportionment of blame.

Inevitably, because so much specialist knowledge is required, the involvement of subject matter experts is a given. Before leaving London, I had been asked if I would assist the AAIB as part of the UK investigation effort and, naturally, I'd said yes.

Whilst we might have had 97 per cent of the facts pertaining to the tragedy, we didn't yet know *why* Concorde had crashed. All we knew was that Air France Flight 4590 had accelerated down the runway, encountered a piece of metal, the tyre had burst and a piece of it had fired up into the underside of the wing, where it had somehow ruptured the Number 5 Tank. Out had poured a torrent of fuel, which had ignited, perhaps from the heat of the afterburners, though possibly, too, from broken or shorted wiring in the undercarriage bay.

After Robin and I left the crash site, we were driven into Paris for a meeting at the DGAC, France's Direction Générale de l'Aviation Civile, their equivalent of our Civil Aviation Authority. The DGAC was located in a tall building with a mirror-glass façade in south-west Paris, just outside the Périphérique ring road. Present were French representatives with a professional interest in the crash, amongst whom was one I recognised: Henri Perrier.

Henri was often described in France as the 'Father of Concorde' – he had been the director of the programme at Aérospatiale, the French prime contractor, now folded and merged into EADS, the European Aeronautic Defence and Space Company. I had met him at Toulouse soon after I'd taken over as chief pilot when we'd initiated the form, fit and function replacement of the computer cards that controlled the aircraft's air intakes. Henri had been in his mid-sixties then, a big, robust man with a full head of hair and a broad, smiling face – at least, five years earlier he had been. Today, he looked tired, greyer and older, but then, I imagined we all did. When Robin and I walked into the conference room, Henri spotted me and waved. There is generally some friendly friction between engineers and pilots, but I'd never sensed that with Henri. I smiled back, took my seat next to Robin, and the meeting began.

I was always a bit cautious about the French side of the Concorde enterprise. Not just because we operated the aircraft slightly differently, or because we were making an OK profit on the back of the aircraft and Air France wasn't; it was to do with something else – call it that infamous Anglo-French divide – that made me think twice about what I said in front of them.

Today, though, there was a palpable sense of unity.

After we'd reviewed some of the salient details of the crash, somebody said we ought to listen to the CVR recording, and instantly there was a drop in temperature. No one wanted to hear anybody's final moments on this, but as one of the few pilots in the room, my observations would be critical.

Someone pushed a laptop to the middle of the table and hit a key.

My French, though far from perfect, was good enough to understand the recording as it opened with Captain Marty, First Officer Marcot, and Flight Engineer Jardinaud discussing their take-off weight and the parameters under which they might have to stop on the runway before their V_1 decision speed.

At 14:40 and 19 seconds (GMT/UTC), having taxied to the threshold, Marty asks how much fuel they'd used.

Jardinaud replies: '800 kilos.'

At 14:42 and 17 seconds, from the tower: 'Air France Four-Five-Nine-Zero, Runway 26 Right, wind zero ninety, eight knots, cleared for take-off.' At which, Marty asks Marcot and Jardinaud if they're ready. They both say yes.

Thirteen seconds later, we heard the clicking of thrust levers as the throttles were pushed forward and a familiar roar – audible as background rumble – as the engines spooled to full thrust.

Twelve seconds after that, Jardinaud says: 'We have four reheats.'

Another eleven seconds and Marcot calls: 'One hundred knots,' A moment later, I heard a very strange sound – a *whoosh*, a sound I'd never heard in an aircraft cockpit before - followed nine seconds later by 'V_1': decision speed, *no going back*.

Approximately six seconds later, at 14:43 and 10 seconds, a yell from Marcot: '*Attention*!'

Then the tower cut in, an urgent male voice: 'Concorde 4590, you have flames, you have flames behind you . . .'

Seven seconds after this, Jardinaud says, urgently: 'Failure Engine Number 2!'

Five seconds later Marty calls for the engine fire procedure, followed by Marcot telling him to watch their speed.

Two seconds later, from Marty, a call to retract the gear and the noise of a selector switch, followed by another sound I recognised: a fire handle being pulled.

For the next ten seconds, there was lots of yelling amongst

them for clarification over the status of the landing-gear retraction, all of this against a cacophony of warning bells and gongs. And then a fire alarm goes off.

Marty to Jardinaud: 'Are you shutting down Engine 2 there?'

'I've shut it down,' the engineer responds.

Marcot calls again to Marty to watch his air speed. Seven seconds later, he says: 'The gear isn't coming up.' This is followed by the *whoop-whoop* of the ground proximity warning system and its chillingly dispassionate robotic instruction to '*pull up, pull up . . .*'

'The air speed,' Marcot yells again. Thirteen seconds later, he says 'Le Bourget' twice.

In the background, scarcely audible above the sound of him struggling to keep control, Marty says something about it being 'too late' against a communication from the tower – a flash message to Charles de Gaulle's fire crews – that Concorde AF4590 was returning to Runway Zero Nine in the opposite direction.

To which, Marcot yells: 'Negative, we're trying for Le Bourget!'

Four seconds later, there's a sharp cry, also from Marcot, it seemed, and then, shockingly, ringing across the conference room, nothing but static.

A number of things later perplexed the investigators when the flight data recorder was analysed, amongst them a decision by Marty to pull back on the stick 15 knots before the prescribed rotation speed.

Anyone who flew Concorde knew that speed was critical and especially at take-off, when you needed to be going as fast as you could.

Having pulled back early, he then allowed the aircraft to drift to the left side of the runway. There would have been a natural tendency for the aircraft to drift left because the fire beneath the left wing would have reduced power on that side; the right engines, meanwhile, were belting out at full power.

We were trained, however, to keep the aircraft straight by using rudder – and Concorde had a very powerful rudder, perfectly capable of keeping it straight under the asymmetric thrust conditions Marty had experienced; rudder power being predicated on a Mach 2 double-engine failure case.

But Marty had allowed the aircraft to drift to the extent it had hit a runway sidelight and was now heading for President Jacques Chirac's 747, which had pulled on to a taxiway at the end of the runway, waiting to cross it.

A piece of folklore sprang up that he had rotated early because the aircraft was pulling to the left and he was trying to avoid Chirac's plane. But the data showed the aircraft was on the centreline, exactly where it should have been, when he lifted the nosewheel at V_R. Only *then* did it start to drift.

The flight data recorder inputs also showed Marty only ever put in 20 degrees of rudder in his bid to correct it – far short of the maximum possible, 30 degrees, the amount that would have been needed to retrieve the situation.

Why?

Had the rudder been limited in some way?

Were the controls damaged?

I remembered the shocking details of the crash of the BEA Trident *Papa India* at the Staines reservoir – how the jump-seat pilot Collins had looked as if he'd been trying to redeploy the slats when he'd been found dead in the cockpit. Had there been some obstruction on the floor of this Concorde that had restricted Marty's input to the pedals?

All of these possible factors were quickly ruled out because the flight data recorder showed that Marty had successfully commanded the full 30 degrees of rudder later in the aircraft's short, disastrous flight.

With no further clues from the FDR, the investigators pressed on.

Having hit the runway light and avoided Chirac's 747, Flight 4590 was now airborne with that horrific rooster-tail of fire streaming behind it – the fire that had been captured so viscerally by several photographers on the ground.

The fire had established itself in an area known as a 'flame-holder' – a volume of disturbed air between the left-hand undercarriage leg and the inner side of the box intake of the Number 2 Engine nacelle.

Flame-holder wasn't a term I'd come across before, but I heard it mentioned several times over the next few days. It described exactly how the fire had been held within a swirling mass of turbulent air inside and around the open undercarriage bay between the inboard engine and the fuselage.

We knew the flame-holder had already begun to cause structural damage, because we had heard it trigger the fire-bell in the cockpit. The cockpit voice recorder mics were so sensitive, they were able to pick up the sounds of switches being activated – right down to the *kinds* of switches.

The fire-bell had been designed to detect a fire *inside* the engine, not an external fire, which set the stage for the second mystery – why the flight engineer, Jardinaud, unilaterally decided to shut down the Number 2 Engine.

Both engines on the left-hand side were what we referred to as 'run back' – they were sucking in smoke, fuel fumes, flames and heat from the external fire, causing what would have amounted to some loss of thrust.

The fire-sensor – the trigger for the fire-bell in the cockpit – was a device called a 'fire loop', a wire running around the engine-bay and designed to melt at a certain temperature, creating a circuit that caused lights to flash and bells to ring. The flame-holder had become so hot, however, that it had melted the fire-loop *inside* the engine bay, triggering the engine fire-warning.

Marty and his crew knew the aircraft was on fire because the tower had told them so.

The combination of these factors – the tower and engine fire warnings – had clearly led Jardinaud to believe the engine was on fire when it wasn't.

Every commercial pilot or flight engineer is trained to annunciate emergencies – especially with engines during the take-off run. An emergency would kick off drills for which they'd all trained. Jardinaud had called, 'Failure engine ... failure Engine 2!' But there was no response from either pilot.

Less than five seconds then went by. Instead of restating the problem, he called 'Shut down Engine 2!' and took the decision to do so on his own.

We trained crews to react methodically in an emergency. A co-pilot's call of 'positive climb' as the wheels left the runway would be followed by the captain calling 'gear-up'. Physically flying the aircraft safely was now the overriding task. To get it stabilised and follow relevant procedures, up to a minimum safe altitude before again agreeing amongst yourselves what the problem was and implementing remedial action. Indeed, the crew manual explicitly said: *'In case of a failure on take-off, no action will be taken before 400 feet AAL, apart from ensuring the track and gear retraction.'* Its implementation would be triggered by the 'handling pilot', in this case

Captain Marty, who would call for the appropriate checklist. Its philosophy was simple: one person flew the aeroplane and gave the instructions and commands; another took the actions; the third monitored them. That was always how it was supposed to be: the rules that had brought it into being were sacrosanct.

What should never have happened was what happened on the flight deck of Flight 4590 – one person, Jardinaud, deciding what needed to be done and taking that action unilaterally. Yet, this, disastrously, is what *had* happened. And not all his assessments had been correct. The Number 2 Engine was recovering from its initial loss of thrust when he shut it down.

And AF4590 wasn't even halfway through its tragically short flight.

In a break in the session, over coffee in one of the rooms off the main conference room, the 'Father of Concorde' came up to me and shook my hand.

'I am sorry, Mike, that we're seeing each other again under such sad circumstances,' he said. 'Under different circumstances, I might have congratulated you on England's *formidable* win in this year's Six Nations.'

I recalled from when we'd met during the computer upgrade programme we'd enjoyed discussions on our shared passion for rugby. France had beaten England that year and Henri was pleased to let me know it. This year, we'd returned the favour and won the championship. 'You still can, Henri,' I said.

Henri suddenly led me to a corner of the room. 'The prosecutor has an *idée fixe*, "an obsession", Mike, about this piece of metal, but you and I know – all sane heads around this table know – that this . . . *énigme* will never be solved by such an approach.'

'So, what do we do about it, Henri?'

'What we do, *mon ami,* is the only thing we can: we keep looking, we keep digging, you using your skills as a *pilote*, me such skills I have as an *ingénieur* and, piece by piece, we get to the truth.'

'Will it do any good?'

'Yes, because it must never happen again,' he said. 'Whatever is in the mind of the *préfet*, finding out what happened must always remain our priority.'

I agreed. I could only imagine how sad it was for him that his creation was not only grounded here, but under such a dark cloud of suspicion. Henri was far too polite to mention it, but I could only speculate, too – knowing of his deep love of country from our rugby conversations – how very upsetting it must have been that Concorde was continuing to fly in the UK, but not in France.

That night, after dinner, I called Chris from my hotel room. I went out of my way not to talk about the crash site or the drama I'd heard unfold on AF4590's flight deck, asking questions about Amy, the cats, the weather back in the UK – anything at all to avoid a discussion of what I had seen and heard that day.

Afterwards, unable to sleep, I got out my laptop and read over the transcript of the desperate final moments of the aircraft based on what the mics had picked up on the flight deck. The background rumble of the Olympuses as the aircraft had started its take-off roll, Jardinaud saying, 'We have four reheats,' Marcot's calmly annunciated 'One hundred knots,' then the *whoosh* sound, Marcot's announcement of

'V$_1$', followed, seconds later, by his warning: '*Attention!*' 'Watch out!'

Although that noise was unlike anything I'd heard on a flight deck, there had been something familiar about it too.

Somewhere in the distance, a police siren wailed. I got up, wandered over to the window, opened it and glanced at the boulevard below and then at the Paris skyline. Beyond the rooftops, I could make out the illuminated upper section of the Eiffel Tower. I rubbed my eyes and walked back to the desk.

Seventeen seconds after V$_R$, Marty had called for the undercarriage to be retracted.

Marcot had selected gear-up, but the gear wouldn't retract.

Gear retraction on Concorde, as any number of walk-arounds pre-take-off always reminded me, was a highly complex process. Because Concorde came in at a high nose-up attitude, it had a very long undercarriage leg – this to obviate the possibility of the engines and tail striking the runway on landing.

One of the first things that had to happen on retraction was for the leg to shorten, so it could fit into the undercarriage bay. It did this by pulling up a shock-absorber built into the leg. The system was so smart that it ensured the undercarriage was capable of being retracted before initiating the sequence.

As well as confirming that the gear legs had been shortened, it checked the nose and main wheels were aligned with the longitudinal axis of the bay before drawing the undercarriage into the belly (thereby avoiding the possibility of the wheels hitting the sides of the undercarriage bays). It also checked that the nose and main undercarriage bay doors were fully open.

Should any of these procedures not occur, the system stopped the retraction.

Concorde's designers had anticipated the possibility of a false or nuisance signal stopping the retraction of an otherwise healthy undercarriage by installing an override switch on the flight deck. Because it was only ever meant to be used in emergencies, the toggle switch in question was covered by a protective flap – and the designers had gone the extra mile by wire-locking the flap in place; so, the chances of it being activated accidentally were, to all intents and purposes, zero. It was, however, a procedure that was listed in the checklist and one that we occasionally had to practise in the sim.

Marcot should have known about the override function because he had been a Concorde simulator flight instructor, just as I had been. And the switch in question was located right in front of him, immediately below the gear retraction lever. He ought to have known, therefore, that the override option allowed them to get around all the protections in place. It was known he hadn't overridden it because the toggle switch was still in its closed, guarded, wire-locked position when they'd recovered it amongst the crash debris.

Had the gear been raised, the whole story might have been different: the drag would have been reduced and the flame-holder destroyed, allowing AF4590 to have maintained its all-important safe climb speed: V_2.

The fact that it had managed to get airborne at all was extraordinary.

Faced with the same issue, a conventional aircraft, at maximum take-off weight, lacking half its normal available power, undercarriage down, would never have been able to get airborne and climb away as Concorde had – a feat made possible only because of the massive levels of thrust available.

Miraculously, Marty had managed to get up to 200 feet with 220 knots on the clock, but beyond this, because of the reduced power and the excessive drag, climbing and/or accelerating further was impossible.

The aeroplane was also becoming more difficult to control.

Soon after clawing its way into the air, there had been an explosion in the 'dry bay' above the Number 2 Engine.

The dry bay, as its name suggested, was an area free of fuel and combustible fluids in which many of the sensitive engine controls and electronics were housed.

Investigators deduced the explosion from the fact that a couple of the bay's access hatches had been found just outside the airport perimeter, blown off by the force of the explosion.

It wasn't just the damage to the aircraft's systems and structures that was causing Marty to struggle. Because of the fuel streaming from the Number 5 Tank, the aircraft's centre of gravity – critical anyway for Concorde on take-off – had shifted significantly aft. Marcot at this point had yelled for Marty to make for Le Bourget, most likely because the aircraft was already pulling to the left.

At one point, as recorded by the FDR, the aircraft rolled 110 degrees left – i.e. 20 degrees past the vertical. It was a testament to Marty's skills that he managed to roll it back again by pulling power back on the Number 3 and 4 Engines – and booting in maximum rudder (which was how investigators knew he'd had 30 degrees of rudder available on take-off). The roll was likely caused by damage to the flight control system as the fire took increasing hold.

But the roll, and Marty's move to correct it, lost the aircraft, already struggling, even more forward air speed.

A conventional aircraft stalls because the nose comes up and air generating lift over the wing breaks away – the lift disappears, the nose falls and the aircraft pitches into the stall.

Concorde's wing was different – it didn't stall. Its lift was generated by those vortices above the wing and their strength was a function of the aircraft's attitude.

As the nose came up, the aircraft flew slower and the vortices grew stronger, generating more lift. But the wing was generating more drag, too.

If the rate of increase in drag was greater than the rate of increase of lift, the drag simply overwhelmed it.

At almost exactly a quarter to five local time, 45 seconds after getting airborne, Flight 4590 simply ran out of energy.

I understood now why the debris field I'd seen had been so contained. Unable to accelerate or climb, it hit the back of the Hotelissimo with its wings level at a mere 100 knots' forward airspeed, but at a rapid rate of descent.

Whilst the flight data and cockpit voice recorders had gone a long way to explaining *what* had happened – they still didn't answer the bigger, more fundamental question: why? Why had it happened? Accident reports from my training and early flying days swam into focus. None of those crashes had been due to a single point failure as was supposedly the case here – the piece of metal on the runway the French were pointing to as the sole cause of the crash. A single point of failure in an air accident rarely, if ever, made sense.

In the small hours, conflating the wail of another siren on the boulevard with a cockpit warning bell, I dreamed I was at the controls of AF4590, wrestling, as Marty had done, to keep the aircraft level, when, as the fire took hold and it had started to come apart, it had flipped beyond the vertical. I heard the screams from the cabin, yelled at the co-pilot that I had this, but the aircraft rolled further, and suddenly I was almost upside down, my gaze locked on the Hotelissimo as it grew and grew in my vision.

I awoke to find myself sitting bolt upright, bathed in sweat, staring at the wall.

The wail of the siren gradually faded. When it had gone, I got up, went to the bathroom and splashed water over my face, grateful that I was heading home that day.

26

Back in the UK, parallel to the investigation, BA was focusing on how the crash would affect its Concorde operations. Air France's grounding of the fleet remained in force. We had cancelled that one flight as a mark of respect, but – pending word from the investigation that this had been anything other than a tragic, freak accident – saw no reason why we shouldn't keep our fleet flying.

That view prevailed if you were a proponent of the aircraft. If you weren't, given the prevalence of the Marmite factor, the crash of Flight 4590 presented a perfect opportunity to remove Concorde from the BA register.

It became obvious fairly early on that Air France was surprised and disappointed that we hadn't stopped our Concorde operations when it had.

There was a lot of pressure on us to stop – to the extent that in the days and weeks after the crash we were constantly being pulled into a series of meetings of the 'interested parties' about what to do. These involved both airlines, both air accident investigation authorities, both sets of regulators, the two governments, and both countries' airframe and engine manufacturers.

The meetings dug into the latest findings about the accident and took place on both sides of the Channel – one week in Paris, the next in London.

If you were a conspiracist, you might have construed France's pressure on us to stop our operations as jealousy that we were flying and they weren't. If you were more generously inclined, you might have decided they were genuinely concerned there was a reason why flying the aircraft was unsafe.

At each meeting, certain postulations were proffered as to why we should stop, none of them valid enough to give me any great cause for concern. My main ally in batting back the barrage of French calls for us to stop flying was a colleague and friend called Jim O'Sullivan, who was BA's chief engineer for Concorde. We weren't under any directive from the senior management to keep the aircraft flying – we just felt there was insufficient evidence to stop operations and that keeping Concorde in the air was the right thing to do.

The meetings dragged on for several weeks at the DGAC and the Ministry of Transport in London. The meetings weren't exactly confrontational, but they were attritional – no one was backing down from their respective positions – and, on top of the emotional toll of the crash itself, I was beginning to feel fatigued by it all.

Then, one day, I got a call out of the blue from Rod Eddington, BA's chief executive.

After congratulating Jim and me on the progress we'd been making, he said: 'Look, Mike, I think you should take a break from all of this – have a few days off.'

I was touched that he was aware that Chris, Amy and I had been due to sail for New York on the day of the crash – even more so that he was now insistent that all three of us should go to New York for the break we never got.

'Listen,' he said, 'I've got tickets for the three of you to fly in comfort to New York the weekend of the 12th/13th of August. It'll do you a world of good. Take them. Go.'

I checked my calendar – mid-August, three weeks after

the crash, was coming up fast. We had nothing on, but I pointed out to Rod that Amy, who was only six, wasn't allowed to travel first class under the terms of our staff travel deal – children had to be twelve years or older to fly there, or on Concorde.

The chief exec stopped me right there. 'These aren't staff tickets, Mike – these are *real* tickets. Real revenue tickets that we're giving you. And they're not Club Class. You're going to get there and back really quickly.'

'You mean?'

'Yes. They're from us to you. *Go.*'

The chief executive had given the three of us full-fare return Concorde tickets to New York. To say I was thrilled, and grateful, was a huge understatement. The break, which couldn't have come soon enough, was scheduled for a time when there were no serious negotiations planned between the two sides. I rang the captain of the Concorde that was scheduled to take us over, explained what was happening, and asked if I might be able to fly the New York leg. He told me he'd be only too delighted.

It was great to be on the flight deck again. After the take-off, Chris and Amy came up to be with me and see the view from the nose. Amy was strapped into the fourth seat and we accelerated through the sound barrier and into our Mach 2 cruise. Throughout it all, however, my daughter was much more interested in the workings of the jump-seat's five-point harness than she was in the fact we were streaking faster than a bullet into the rising sun.

When we arrived in New York, my colleagues on the PR side did us proud. One of the things that Amy said she really wanted to do was to ride in a stretch limo into Manhattan, which the PR people had organised. We were dropped off at the Helmsley Palace Hotel and were given an amazing room overlooking St Patrick's Cathedral. Amy became fascinated by the white carpet, which was ankle-deep, even for us grown-ups.

Before I left, I had mentioned in passing to someone on our negotiation team that I was going away for a long weekend. I'd thought no more about it. Whilst aware that 15th August was a prominent Roman Catholic holiday, the Feast of the Assumption of the Virgin Mary, I was not aware that Jim O'Sullivan, who was Catholic, had arranged to travel back to Ireland to see family and friends over that weekend. Nor did he have any idea that I was going to be in New York.

It was pretty obvious in hindsight that elements of the working group who were opposed to the continuation of Concorde operations – part of the Marmite anti-set – had seen a window of opportunity: a period when both Jim and I would be away; we being the two main proponents of continued flight operations. Had I been a little less weary, I might have seen it coming, but I didn't.

On the Monday, I got a call from Jim to say that the DGAC and the BEA had just published some preliminary conclusions on the causes of the crash – it had been down to a single point of failure, they said: all caused by the wearstrip on the runway.

The DGAC maintained that Concorde had been prone to this kind of risk all along and was, thus, inherently flawed. The DGAC was, as a result, removing its Certificate of Airworthiness. The UK's CAA was bound to follow suit.

As if this wasn't bad enough, BA had accepted the arguments and had grounded our Concorde fleet on the spot, even though the removal of the certificate didn't come into effect for a further day. Had Jim or I been there to defend

Concorde, the result would almost certainly have been different.

But we hadn't – the conspiracy had played out while the loyal guard had been away.

Concorde had fallen foul of what amounted to a good, old-fashioned *putsch*. Either that, or it was one heck of a coincidence.

The *coup d'état* meant I needed to get back to London fast, which is when the ironies kicked in and compounded.

I had been booked on a Concorde flight first thing the next morning, which would have been perfect – it was, after all, the kind of problem for which the aeroplane had been built and procured. But the grounding meant the flight no longer existed.

The PR people in New York swung into gear again and managed to get Chris, Amy and me on to the first 747 flight – a daylight flight – back to London.

Once on board, I got a close-up view of what it meant to withdraw an entire aircraft type from service overnight. The 747 was packed – a sizeable proportion of the passengers, like me, had come from the cancelled Concorde flight and the mood amongst a good many of them was febrile. One man, upon learning I was a Concorde captain, took great umbrage from the fact I was on board. He eventually calmed down when I convinced him I was on his side – that I had to get back to London as soon as possible so I could work with my colleagues to rectify the situation and get Concorde fly-ing again.

Settling back into my seat, the magnitude of the task ahead began to sink in.

The priority still was to find out what had caused the accident.

The argument put forward by the DGAC was crazy. For an aircraft to suffer from a single-point failure was absurd – I didn't buy it for a second and nor did Jim. Single-point failure accidents in aviation were vanishingly rare.

The accidents with which I was familiar – the 1-11 prototype that had super-stalled over Wiltshire; the BOAC 707 whose engine had fallen off after take-off from Heathrow; the BEA Trident that had crashed into the side of the George VI reservoir near Staines; the Tu-144 'Concordski' at the Paris Air Show; even Bill Weaver's SR-71 Blackbird – they had all come about owing to a causal chain of events. That was the way almost all accidents happened.

Concorde's failure, I told myself, had to be no different. I knew my aeroplane. More to the point, Jim did. The aircraft had been robustly designed. The tragedy at Gonesse had almost certainly been avoidable – most air accidents were. But the DGAC verdict strongly implied that what had happened had been *inevitable*, because of an inherent flaw in the aircraft.

As with all *putsches* and *coups*, this one was mired in politics. It was fairly obvious Air France wouldn't consider it a disaster at all if their aircraft never flew again. There were a number of reasons why. First, they had never made Concorde the commercial success it had been at BA. Primarily, this was because they had never divorced themselves from the arrangement under which their government had committed to cover their losses and take all their profits – a very different situation from that which had existed at BA after the Lord King 'deal' with the Thatcher government in 1984.

Air France's books registered Concorde as a massive loss – and nothing looked likely to change that.

To make matters worse, twenty years after BA had come off the UK government's books, the French flag-carrier was coming up for privatisation. One of the ways to rid itself of the Concorde loss was to stop operating it – to write it off the books.

Third, international law governing corporate liability was changing. Going forward, directors could be held accountable for major accidents that had taken place in their companies on their watch. Putting as much distance as possible between the airline and the crash would be seen by some as a smart move.

All told, Concorde had embarrassed Air France – and because of the ties that bound the airline to the French government, it might not have been a stretch to say it had embarrassed France. The mood music coming out of Paris said that very few politicians on the other side of the Channel would be put out either if she never flew again.

That, I told myself, was their affair. At BA, things were very different. And yet, some of the same Machiavellian undercurrents prevailed. It wouldn't take much for the Marmite *putsch* that had led to the grounding becoming permanent.

My capacity to fight was limited, but I decided to give it everything I'd got. I was Chief Concorde Pilot in name, but my actual title was still Manager Technical/Manager Training – Concorde. Both roles were what we called *alpha*-grade, senior management, but, in the pecking order of things, still relatively junior – there were four whole management layers above me (all the way to *epsilon*, the Greek 'e') before I was able to knock on the door of the chief exec. It had never occurred to me to seek out senior BA permission for the strategy Jim and I had embarked upon; it had just seemed the right thing to do. Jim felt likewise. Whilst my determination was based primarily on passion, Jim always said his was based on logic – there was no reason, he said, to stop the aircraft flying.

My first action, then, on landing back at Heathrow was to seek Jim out – he having returned from Ireland a day earlier. I found him in his office in a sombre mood.

'Listen,' I said, 'we've got even more of an uphill battle than we had before. Are you still up for it?'

'More than ever,' he replied.

'To get her in the air again we're going to have to persuade a lot of people about the investment levels that are going to be necessary.'

He nodded. 'I know.'

'And it looks like it's going to be just you and me leading the charge.' I paused. 'We're only going to succeed by demonstrating some real passion.'

'I don't do passion, Mike,' he said. 'I do logic.'

'OK,' I said, 'I'll do the passion, you do the logic.' We shook on it.

What we'd embarked upon was not against the wishes of the senior management – had it been, of course, they'd have stopped us. But it wasn't with their explicit direction either. We were doing what we were doing, it felt to both of us, because it was right, and something we had to do.

27

From the beginning we were presented with a number of challenges. Hot on the heels of the BA Marmite anti-brigade, the biggest was the task of devising a set of mods that would meet the outcome of the accident investigation, which was ongoing. The BEA's preliminary investigation would be followed by an interim report before, finally, being signed off and sealed by a main report that mightn't be finished for another 18 months. The plan to get Concorde back in the air wasn't contingent on the publication of the final report but would need to take in all the various recommendations that would emerge from the investigation along the way.

The other big challenge was the French, whom we would need to bring along with us.

If they dug their heels in and said they weren't going to recommence Concorde operations at Air France, then it would be game-over.

This was because the millions of pounds that we paid annually to the manufacturers in support costs – before Concorde moved an inch off the ramp – was predicated on there being two operators: Air France and us. If they stopped operating, all the infrastructure bills would fall on us – a massive increase in our costs and one, I suspected, that would be totally unsustainable. And even though Concorde was meant to be protected from unilateral withdrawal by one of the two partner nations – an agreement that went back to the signing of the original international treaty in the mists of the

1960s – we couldn't exclude the possibility that France would use the shock of the accident to wriggle out of it. So, keeping Air France motivated, and on-side, would be vital.

Then there were the practicalities of doing the mods to consider – not so much the practicalities of designing and building them, which we had great confidence would be accomplished, but the process by which they would be implemented and tested. The manufacturers – BAE Systems and EADS – no longer had a Concorde at their disposal on which they could install and test the new systems that would satisfy the authorities' demands for a flawless reintroduction to service. The only aircraft that remained outside of airline service was *Delta Golf*, which BA owned and had once been a test aircraft. But *Delta Golf* hadn't flown for years and was practically incapable of being made airworthy – to have done so would have cost a fortune nobody had and would have taken too long. From the get-go, I said to anyone who would listen that we had to get the aeroplane back into the air within a year; if we didn't, it wouldn't happen. And, ideally, I said, it would be better if it were three months.

My reasoning was as follows. I could stand my flight crew down for three months; Jim could do the same with his engineers. We could bear the loss of them sitting down, doing nothing for that length of time. If it increased to six months, things would begin to get marginal. I would have to redeploy some of my pilots onto other aircraft types and Jim would have to do the same with his engineers. There was a breakpoint – the point at which the cost of retraining them eclipsed the cost of them sitting around, cooling their heels. It might even be possible to stretch that period to twelve months. But that would be the absolute limit. By then, with aircrew and engineers installed on other types, the experience base we'd built up over twenty-four years from 1976 would have been diluted to a point where it would probably be impossible to reinstate it.

We had a year, maximum, I reiterated, and then that would be it: no more Concorde.

The year became a mantra amongst all those who wanted to see the aircraft get back into service. It drove everything we did. The aircraft ceased operations on the 14th August 2000, three weeks after the crash. If she wasn't back in the air in her fully modified, approved form by the 14th August 2001, a year later, we could all effectively go home.

So, the fact that we didn't have a test aircraft was an issue. If the manufacturers didn't have one, and *Delta Golf*, our test aircraft, was beyond hope of becoming airworthy in time, it was clear that the mods would need to be installed and approved on one of our fleet aircraft, which would become the *de facto* test vehicle. But this wasn't as simple as it sounded. The qualifications needed to conduct the sorts of test flights we were talking about were vested in a very small cadre of pilots – they were CAA-approved test pilots – and there were none employed by BA. The CAA had one test pilot who was Concorde qualified – Jock Reid. In his role, he sometimes flew our aircraft, with passengers, on our routes as a co-pilot alongside a BA training captain. In that guise, he was covered by all the necessary insurances that BA had to hold. I knew Jock well and was very comfortable with his being closely involved. Plus, it seemed to be the 'politically correct' thing to do. Better 'inside the tent' than out. The remaining conundrum revolved around the fact that the CAA had the qualifications, the regulatory authorities and the testing expertise and we had the aircraft, the insurances and the operating expertise. In the end, the CAA came up with a compromise that saw them issuing their test pilot license to me and the equivalent to the BA flight engineers who were going to fly the test aeroplane. At the same time, we agreed that we would provide an aircraft from the fleet to the manufacturers for the duration of the modifications and for the test at zero cost. With the ticking of the clock a constant reminder of what we needed to do in precious little time, we were off.

It was good that our hand was on the tiller – better still, that the tiller itself belonged to us – otherwise I'm not sure the modification programme would ever have got under way. The pressure from the nay-sayers within BA was constant; and were it not for the fact that we were now in the driving seat, the nay-sayers would have put the whole thing on the 'too difficult pile' – which is where it would have stayed. The blame would then have been placed on all the other players with a hand in the accident investigation and the modification effort, and the Marmite anti-brigade could have washed their hands of the affair, safe in the knowledge they would have been absolved of blame – the backlash that would have come from eradicating an icon from the skies.

But because we were in control, the aeroplane's destiny – for good or ill – was in our hands.

We knew that to bring Concorde back into service we would need to mount a very carefully calibrated PR campaign, with two main constituents to convince in parallel to a successful outcome of the engineering works.

The first was the public and its perception of the safety of the aircraft.

The second was her safety perception from the viewpoint of our customers – two things that were linked, but different.

There was also, as a subset of them both, an internal PR campaign we needed to drive because, clearly, if we didn't have the support of a majority in BA and the people at the coalface – all of the people who were responsible for delivering Concorde as a service to the customer – then, again, none of this was going to work. And this extended from the flight and cabin crew all the way to the baggage loaders. Anyone who questioned me on this, I invited them to consider the damage that could be done by a baggage handler, enquiring in a friendly way where you were headed when you dropped your bags off at the curbside, and you being greeted with a pained face and sucking of teeth when you told them you were on Concorde. *'Oh, dear, are you sure you want to do that? I mean, wouldn't you rather rebook on a 747?'*

So, I began running regular briefings to and with anyone who would be a touch-point for the aircraft when she came back into service, ensuring that everybody was kept up to speed on where we were and progress with the modification programme; this so that they could address with confidence any query that came their way – be it from a prospective customer, a supplier, a work colleague or, simply, from a mate down the pub. This linked, too, to getting 'public opinion' on our side. The effort led by me in Operations was mirrored by a similar, internally focused campaign mounted by Jim in Engineering.

We would only have one shot at this, so everything we did had to count.

And then there were the customers. We had been operating long enough to know our customer base pretty well. Of the 80 per cent who were business passengers, 80 per cent of *them* were repeat customers. When the time came, we would pull them in, too, and let them know what was happening. In rebuilding confidence around Concorde, transparency would be everything. But across the Channel, where the criminal investigation was proceeding apace alongside the race to put together the pieces of the crash investigation – metaphorically and literally – transparency was a commodity that wasn't always in the abundance of supply we'd wished, or even hoped, for.

28

The pieces had been gathered into a hangar in a discreet corner of Le Bourget, the airfield Captain Marty and his crew had desperately been trying to make for when their luck, and that of all on board, had tragically run out.

Hangars have a distinctive smell, the prime components of which (as associated by me, at least) are oil, glues and solvents. But on my first visit to Le Bourget, they were joined by an unwelcome *over-odour* of burnt metal and plastics. When I first saw the pieces on the floor – twisted, charred and barely recognisable as an aeroplane, let alone the aeroplane I knew – the shock and grief I'd experienced while standing on the fire-escape stairs at the hotel in Gonesse might have resurfaced had it not been for a conscious effort on my part, as well the team's, to treat our role in the investigation with the professional detachment it deserved; the only way that we knew to get the job done.

The legal agreement that had been thrashed out between our two countries' investigation authorities should have given us full access to everything the French turned up on and off the crash site. But right from the start, it was clear that the prefect, the magistrate whose authority was Napoleonic in its grip on the process, was not so keen to share certain pieces of evidence with us.

Chief amongst these was 'Exhibit A' (in its metaphorical sense), the titanium wear-strip that had fallen off the thrust-reverser 'clam-shell' door of the Number 3 Engine of the

Continental Airlines' DC-10 that had taken off from the same runway as Flight 4590 a few minutes previously.

We were only ever given the most cursory opportunity to examine this and other critical components; the way in which the wear-strip had come to be guarded being such that we – the Brits on the investigative team – began to joke about our suspicions that the prefect was keeping the damn thing under his bed.

Furthermore, whenever we paid particular attention to a component or a set of components in the hangar at Le Bourget – which was always, and rightly, guarded by an ever-present *gendarme* or two – we began to notice by the following morning that the component/s had disappeared from the floor.

When we asked, we were politely informed that the prefect had found it necessary to impound them.

So familiar did this pattern become that we developed a set of hand signals to indicate amongst us only, or so we thought, that we had found some article of significance. But pretty soon, they rumbled this, too, and we became forced, like schoolboys behind the bike sheds, to hold whispered, huddled conversations about our daily observations back at our hotel.

The tyre destruction was obviously central to the investigation. It had been known from those several incidents involving burst tyres back in the early days of service, and that strongly worded report from the US air accident authorities, NTSB, that Concorde had a susceptibility to fuel leaks from bits of rubber bursting through the wing. The punctures in those incidents had been small – and the remedy, we all thought, had been robustly simple: we'd stopped the early-days practice of using retreads and started using new tyres instead; this in addition to the flat-tyre detection system that alerted the crew to an issue.

At BA, we had also installed that change to the wheel structure – the cable that stopped the water deflector flying off – but Air France hadn't. By early August, around a week after the crash, the media had picked up on this, forcing Air France to issue strenuous denials that this had had anything to do with it.

Several months after the crash, I was back in the hangar as part of a team of experts from BA, Air France, EADS and BAE Systems to review a second phase of reconstruction of the wreckage. This had focused particularly on the left wing and the point where it was calculated the strip of rubber from the scalped tyre tread had fired into its underside. Gazing at what the investigators had been able to recover was a distressing reminder of just how little of the aircraft remained. The pieces had been laid flat on two areas of the hangar representing the upper and lower wing surfaces. Unless I'd been told what I was actually looking at, I might have deduced that I was in a scrapyard.

Nothing remained at all of the landing gear wells or the areas of wing surrounding them and the same could almost be said of the entire left wing.

There were precious few parts either from Tank Number 5, the tank immediately above the wheel that had run over the metal strip, where the leaked fuel was suspected of coming from. Three pieces, however, had been identified – one found on the runway, the other two at the accident site. The piece on the runway was the piece that the investigators had focused on because its presence there was not immediately understood. Why had it been ejected at the moment it had been struck by the piece of rubber from the tyre?

When we'd asked to see this piece, the authorities told us it wasn't possible. I don't recall why – it may have been down to another frequently uttered catch-all – that it was undergoing '*une analyse*'. But one of our team, a guy from the AAIB, had spoken to his French counterpart who'd revealed the piece, which was more or less square – roughly 30 cm by 30 cm – had shown unusual 'deformations', their conclusion being that Tank Number 5 had not been punctured from the outside in, but rather it had burst from the inside out.

How, I'd asked our AAIB metals expert.

'I don't know,' he said. 'We're going to need to give it some thought.'

A few days later, back in the UK, I was at BAE Systems's Filton facility flying the sim, evaluating Concorde's ability to climb away on two engines under the conditions it had experienced on the day of the crash – all underwritten by the maximum take-off weight. After I'd finished, I ran into an engineer I'd known for a long time – I'll call him Tom. He asked me how things were going on the other side of the Channel. I told him about the piece found on the runway from Tank Number 5 and the French conclusion that it had burst from the inside.

Tom said: 'You might want to ask, then, if the tanks were overloaded.'

'But . . .'

'Yeah. If they were overloaded and a piece of something hit the wing, the impact could have set up a shock wave in the tank causing it to blow. But such a thing would only be possible if that tank had been filled to capacity.'

I wasn't due back in Paris for a while so passed on what he'd told me.

Several days later, I got a call from the same AAIB metals expert, who was phoning from a callbox close to the hotel. He sounded excited, breathless. 'I ran that tip about over-loading past our French friends and got a very interesting reaction.' He paused. 'More of an *over*-reaction, actually.'

'What do you mean?'

'*C'était impossible*,' he said sarcastically. 'No way could the tanks possibly have been over-filled. Air France wouldn't ever do such a thing.'

'Then how do they account for the tank rupturing from the inside?'

'Short of it being filled to capacity, the theory they're going with here is that as Concorde accelerated down the runway, the fuel surged to the back of the tank, where it was *effectively* full. The empty part of the tank was at the front – some distance, they calculate, from the point where the rubber struck the underside of the wing; at which point, they're saying, the tank was, to all intents and purposes, full. *Ha!*'

A fuel tank is either full or it isn't.

To accept a massive shock wave inside Tank Number 5 had caused it to burst from the inside, it had to have been 100 per cent full. No tank was ever supposed to be 100 per cent full for this very reason – the specified, don't-exceed level in Concorde's tanks was 95 per cent. It began to dawn on me that lurking somewhere in all the fuel data lay the real and tragic story of Air France Flight 4590.

Concorde, as we all knew, was limited on range on certain routes, and so, not unreasonably, you'd want to get as much fuel on board as you could – as you were *allowed*.

Towards the end of the refuelling process, all the tanks in the aircraft – the thirteen tanks fell into four categories: four engine feeder-tanks, four main transfer tanks, two wingtip 'A' tanks and three tanks used for trimming in flight – would initially be filled to the maximum level permitted: 95 per cent.

All, except Tank 11 beneath the fin, one of the three trim tanks, which, at this stage, was usually filled to just under 50 per cent, setting the aircraft's C of G just forward of 54 per cent.

'If you wanted to put *more* fuel on board after all the tanks had been filled to their maximum permitted capacity,' I told a meeting of my fellow investigators several days after the call I'd received from that Parisian callbox, 'the only place you could do it was Tank 11.'

The effect of this, however, would be to shift Concorde's C of G aft – and the C of G, I reminded them, needed to be religiously set at 54 per cent for take-off.

The more Tank 11 filled, the further aft C of G would move. For the sake of argument, I invited them to envisage a scenario in which, on the day of the tragedy, it might have been moved to as much as 54.4 per cent aft.

Because the C of G *had* to be 54 per cent on take-off, under the 54.4 scenario, a chunk of fuel would have to be moved forward for take-off. But all the other tanks were full to 95 per cent – the maximum allowable – so you'd be unable to transfer any fuel to them until they had some spare capacity.

Capacity would only materialise when you'd burnt off some fuel.

'So, let's say,' I told my colleagues, 'the amount of fuel you need to move to the forward tanks to rebalance the C of G from 54.4 per cent to 54 per cent is 1,300 kilograms. That 1,300 kilos can only be moved when 1,300 kilos have been burnt off from the feeder tanks.' What followed, I said, was a little like one of those kids' puzzles where you slid squares around a board – where you had little room for manoeuvre because you only had one missing space.

'Thirteen hundred kilos is extra fuel and is precious,' I said. 'But there's a moment when there's no point putting *more* fuel in Tank 11 because your C of G will shift even further aft, requiring you to burn more fuel before take-off.'

You hit a point, in other words, where it became self-defeating.

I then told them how, in my early days as Chief Concorde Pilot, I'd become aware of an unapproved procedure amongst some of our BA crews that had seen the aircraft's tanks, on occasion, filled beyond 95 per cent.

At BA, we'd learnt that there was a margin – ever such a slight one – to that magic C of G figure of 54 per cent. The margin was +/- 0.2 per cent, which was there to accommodate instrumentation errors. In the fuel transfer process, the computer was supposed to be used as a verification tool only. The primary mode of obtaining 54 per cent was to pump fuel around using the data that came from the load sheet. Its calculations would determine how much weight – in other words *fuel* – had to be moved forward to shift the C of G from its pre-engine start figure of 54.4 per cent to the take-off figure of 54.

The load sheet stated the amount of fuel that needed to be moved forward from Tank 11 and the computer would then verify it. The flight engineer would monitor and oversee the process, check against the computer's own figure and, *bingo*, if the two matched, you knew (with something approaching 100 per cent confidence) you were good to go.

'If the computer said 53.8 per cent or 54.2 per cent, you

were still good to go, because that was the margin of error.'
In my Concorde career, I added, I'd never seen that much of
a margin – very occasionally, a 53.9 or a 54.1 – because the
system was *that* good.

But it was technically possible to work the system the
other way.

Instead of using the computer as a verification tool for
the amount of fuel that needed to be transferred under the
load sheet calculation, the fuel transfer could be manually
stopped when the computer read 54.2 per cent.

In other words, you could fill it deliberately to the upper
limit of the margin of error.

The heavier an aircraft, the more fuel it burns going from
A to B.

On a conventional aircraft, this is a relatively low additional amount – about 15 per cent, or 150 kg for every extra
1,000 kg (or one tonne) loaded. It was the nature of 'the
Beast' – the lore of Concorde fuelling – that for every tonne
of fuel you put on board, you burnt half that amount flying
it to your destination.

So, when you put on a tonne and flew across the Atlantic,
because the aircraft weighed a tonne more, it would burn
half a tonne to carry that tonne. When you got to the other
end, logically, you'd only have half of that tonne left.

But the maths and physics worked the other way round as
well.

If you had a marginal operation – as Flight 4590 was – and
you put as much fuel on as you possibly could, and you found
yourself a tonne short of what you needed, the science said
that, because for every tonne of fuel you put on board, you
burnt half of on route, then you would be 500 kg 'short'
when you got there.

An extra tonne on board at take-off from Paris, therefore, would be worth 2 tonnes when the aircraft got to New York. Those extra 2 tonnes would have been a big incentive to fudge the system on what was an extremely marginal flight. Ordinarily, I would have said a scenario like this was highly unlikely, but we'd just been informed that all copies of the load sheet – the document that would have given us all this information – were missing.

29

In the weeks and months after the crash, BA was having to react on multiple fronts. There was the investigation in Paris. There was the public reaction to manage. There was the rearguard action fought against the nay-sayers in BA by those of us who wanted to see Concorde back in the air; and there was the mod programme.

In the absence of a final report from the air accident investigators, but understanding the 'root causes', we collectively decided to attack the 'causal chain' in three places: the tyres; the ruptured wing tanks; and the ignited fuel.

The tyre, we now knew, had been destroyed in a unique way. All aircraft certification involves tests on tyres that might fail – what you had to do was demonstrate as manufacturers that the aircraft structure could resist impact from a piece of tyre 'squared by the width of the tread'. Concorde's tread was 33 cm wide, so we were talking about a piece of rubber 33 cm² that, in trials, had been fired at the underside of the aircraft at representative velocities – the kinds of speeds you'd encounter on a tyre failure at take-off or on landing (the more critical of the two being take-off where speeds, weights and forces were higher). This magnitude of test was predicated on what was called a 'straightforward fail' – a puncture/carcass/tread failure.

This tyre failure wasn't like that.

The piece of metal that had detached from the Continental

Airlines DC-10 was a titanium wear-strip from the right engine thrust-reverser mechanism.

The fact that it was titanium was unusual. A repair had been carried out on the DC-10 some months earlier, when its predecessor part had done what it was supposed to do. A wear-strip attaches to the underside of a thrust-reverser bucket to wear away should the underside of the bucket scrape the runway on touchdown. Wear-strips are normally made of strong, high-grade aluminium. But when the original piece wore out – it had been replaced during recent routine maintenance – the maintenance team had not had any durable aluminium available, so had replaced it with a titanium component instead.

Titanium is tougher and lighter and meets the spec required by the maintenance manual. It wasn't the normal material for such a repair, but it was legitimate.

The piece in question was roughly 45 cm long, 3 cm wide and 3 mm thick and had been bent into a peculiar, twisted L-shape. It was this piece – the accident's metaphorical 'Exhibit A' – that had been so guarded by the prefect that we'd never actually examined it. But we'd seen pictures; some in sufficient detail that the UK investigation team had been able to mock-up a replica.

Using the replica in trials, it became clear that the way it presented to the tyre was critical.

Placed on the ground one way, nothing happened. The tyre just ran over it.

But placed another way, it was lethal. The way it presented was down to its peculiar twisted, bent shape. Expressed slightly differently, you could say chance had dictated the way it fell that day – it had been down to the roll of a dice: pure fate.

Be that as it may, when Flight 4590's front-inner tyre on the left-hand leg ran over it, the titanium 'scalped' the tread. Instead of the 33 cm² piece of rubber that met the specs and safety requirements flying up, away came a piece 33 cm wide – the width of the tread but more than a metre long and weighing approximately 4.5 kilos – propelled by a wheel doing about 205 mph and given the added energy of rotation, a unique failure mode, unknown – as far as I am aware – in civil aviation.

The rubber strip hit the underside of the aircraft in an area containing a wing-tank.

On Concorde, the wing *was* the tank. There was no bladder inside the wing containing the fuel. The fuel was pumped directly into a sealed section of the wing itself. Each of the thirteen tanks distributed throughout the aircraft was fundamentally a part of the structure. There was no internal lining, even.

The piece of rubber hit squarely in the area of Tank Number 5, above the destroyed tyre.

The reason for there being a supposed air gap in the tank was to allow for expansion and contraction – be it on account of the medium/long-term effects of temperature change; or the acute, short-term effect of an impact like this.

Because, as we now began to suspect, the tank had been 100 per cent full, there had been nowhere for the energy imparted by the rubber to go – no means for it to be dissipated within the fuel. The energy, therefore, bounced around inside the tank until it forced out a section of the skin roughly 32 by 32 cm² – about the same area as the front page of a tabloid newspaper.

Not a puncture from the outside, but a rupture from the inside.

This was again unprecedented physics in civil aviation. It had been known to the military, which was aware of this type of tank destruction under certain kinds of combat-induced impacts from particular kinds of munitions.

The next thing that happened – the next link in the causal chain – was the 100 litres of fuel per second that poured out of the hole caused by the rupture in the tank.

I recalled the great BBC TV science journalist, James Burke, reporting for the *Tomorrow's World* programme many years earlier standing in a pool of kerosene – jet fuel – while he'd held a lighted match. The climax of the demonstration, which may have even been conducted live (I can't now remember), was when Burke dropped the match . . . and nothing had happened. Burke had made his point well. Jet fuel has a high flash point and is hard to light. It takes the complex turbomachinery of a jet engine to do this.

One of the other conditions under which jet fuel will ignite, however, is rapid vaporisation – so-called 'misting' of the fuel, which you will get when 100 litres per second pours out of a hole, such as the one made in Concorde's wing. The effect is like a fire-hose – at the edges of the stream, fuel vaporises into the surrounding air, where it is highly combustible.

Next, you needed an ignition source.

In the crash investigation, there was a bitter debate about what this might have been.

The French doggedly insisted that the vapour cloud had been lit by the reheats as it had flowed into the path of the hot, afterburning gases of the two Olympus engines under the port wing.

The British, who had design responsibility for Concorde's fire-detection and suppression management systems, proved to their satisfaction – and, incidentally, to mine – that this was impossible because there were soot deposits on the runway coincident with the initial point at which the fuel first burst into flame.

Further up the runway, there were fuel deposits marking the point at which the unlit fuel had gushed from the hole in the wing.

Knowing certain parameters – the aircraft's speed, wind direction and other factors – the British investigators were able to calculate the precise moment the fuel had first exited the tank and the point of ignition. They knew, by working out the number of milliseconds it had taken for the fuel to ignite, that it could not have been from the reheats; this because of the time it would have taken for the flame-front to work its way back up the fuel stream to the seat of the flame-holder – the area of turbulent air between the undercarriage and the inner engine intake box.

They calculated that the time interval between the fuel hitting the reheats and the fire's grip on the flame-holder had to be down to something else.

The French refused to be convinced and, in the end, a compromise was reached: the fire, they said, was most likely started by sparking from the wiring in the undercarriage bay – specifically, they said, from wiring taking electricity to the fans that cooled the brakes, which could get very hot.

From this analysis, we could break the chain in three places: the blown tyre, the hole in the tank and the wiring.

To get Concorde back in the air, then, this was what the mod effort would need to address.

I learnt that Michelin had already produced a new tyre for Concorde, called 'Near-Zero Growth'. NZG, as its name suggested, which could be pumped up to full pressure and hardly expand in the process – quite an achievement when you take into account what happens to a bicycle tyre or a balloon that's over-pumped. NZG was made with Kevlar aramid fibre reinforcement that wasn't only resistant to foreign-object impact but, were it to blow, would disintegrate into smaller pieces than the existing tyres.

The most challenging part of the programme – and the most integral to the cause of the crash – was addressing the issue that had led to the hole in Tank 5. This was to be achieved, first, by fixing the tanks with linings that would minimise the shock wave impact that had led to the rupture and, second, act to plug the hole should such an event occur.

Only certain tanks were susceptible to this damage, so these were the ones addressed, all of them on the aircraft's underside.

The liners were made of carbon-composite Kevlar saturated with a specialised sealant, which also helped to mould the liners to the bottom of the tanks. The design, manufacture and fitting of these liners, which all had to be custom-made, was the long-lead item in the mod programme.

More than a hundred linings would need to be fitted onto each aircraft, each designed to be fixed to the bottom of the tanks through the use of special brackets.

Because each Concorde was itself bespoke – all were, in effect, hand-built, and very slightly different – this, we knew, would be the most complex and time-consuming task.

We also knew that the liners would add weight and their presence in the tanks would act to displace some fuel, potentially reducing Concorde's range.

But this would be a small price to pay for the safety assurances provided – and, crucially, we knew that the liners wouldn't significantly alter the aeroplane's all-important centre of gravity profile. We resolved to address the weight issue later.

The final significant part of the programme was completely armouring the wiring in the undercarriage (the looms were already partly armoured) to increase its resistance to impact damage.

We also agreed that the operating procedure needed to be changed so that we never again took off with that particular 115-volt brake cooling fan circuit 'live'.

With a few other minor mods – the removal from the undercarriage of a water deflector retention cable, a slight reprofiling of the deflector itself, new anti-skid protocols, a system to ensure that the flat-tyre detection system was 'active' on take-off and ensuring that power to the brake cooling fans was off prior to take-off – the programme was agreed, and the work started. By undertaking it, by breaking the chain in three places, we were ensuring that this kind of accident would never happen again, and it came with a cost implication: £14 million for the BA Concorde fleet alone. The condition we placed on the spend was my milestone for completion and the aeroplane's service re-entry – no more than a year on from the date of the accident.

The team that had designed the mod and come up with the fix was a mixed crew of Airbus UK and Airbus France, regulatory technicians from the CAA and the DGAC, and engineers from BA and Air France. The development was a coordinated effort – the installations themselves would be down to each airline's engineering team.

When it came to testing what they had come up with, the focus of the effort shifted to the French equivalent of the British military aircraft testing facility at Boscombe Down – a secretive airfield in southern France called Istres.

30

The Istres flight test centre, known in France as the Istres *Centre d'Essais en Vol*, was a large air base in a marshy region between Marseilles and Montpellier, approximately fifteen kilometres from the Mediterranean coast. Its five-kilometre runway, the longest in Europe, was so long it had been offered by the French to NASA as an emergency landing strip for the space shuttle.

Spooky things were said to go on at Istres, due to the occasional presence of a US U-2 spyplane – a deployment that coincided with periods of tension and conflict.

One half of Istres was given over to the military activities of the French Air Force, the other to the flight test of aircraft that the air force was upgrading or bringing into service. In this part of the base, France's two main aerospace manufacturers, Dassault and EADS, maintained permanent test facilities – and it was to these that Concorde F-BVFB taxied after landing from Paris on 18th January 2001, the first time she'd flown since the accident. The fifty-minute flight from Charles de Gaulle had been made possible by a special Certificate of Airworthiness issued by the French authorities that allowed the trial to take place.

Dozens of fans of the aircraft had gathered in the freezing cold at the end of CDG's runway to watch her depart. Some had hung a banner on the perimeter fence declaring: 'CONCORDE ON T'AIME'. *Concorde, we love you.*

Four hundred miles south, where the weather was

appreciated for its warmth, I arrived at Istres for the first of two key sets of trials to a marked chill in the air. This could not stop my excitement at seeing F-BVFB on the apron and technicians swarming around her as she was readied for the day's test.

The first set of trials involved the simulation of fuel leaks from the wing tanks as the aircraft travelled at speed down the runway.

F-BVFB had been marked with hundreds of small tufts on her left underside where the flame-holder had taken root, next to the wheel well and the inboard wall of the Number 2 Engine.

These would enable engineers to see how leaks from the Number 5 Fuel Tank followed the airflow around the engines – this thanks to dyed water sprayed from the leading edge into the intakes and thence left to stream back to the exhausts.

The cabin had been stripped of seats and filled instead with sensors and instruments to record the flow patterns of the yellow dye as the aircraft roared up and down the runway. The point of the tests was to understand how leaked fuel ingested into the intakes affected the performance of the Olympus 593s on take-off as well as how the reheats, under certain conditions, might conceivably ignite the fuel as it streamed back over the wings to the exhausts.

The tests went so well that they allowed us to proceed to the next part of the 'causal chain' – the fitting and testing of Kevlar linings in the fuel tanks of another test aircraft – our very own G-BOAF *Alpha Foxtrot*, which would be readied for flight trials during the early summer.

In the meantime, trials involving another French aircraft, F-BTSD, were set for April and May 2001, again at Istres. Together with the flight trials involving *Alpha Foxtrot*, this would allow officials on both sides of the Channel to make a final determination about returning the two fleets to service.

It was a glorious sunny day when I flew into Istres at the beginning of May for the second of the two trials planned to take place there – the testing of the burst-resistance Near-Zero Growth tyres designed and developed by Michelin. Confidence in these tyres was high following earlier tests involving a tractor fitted with one of them. The NZG tyre had been weighed down under 20 tonnes of breeze-blocks as the vehicle had driven over a strip of metal like the piece that had scalped AF4590's left inboard wheel tyre. The results had been sufficiently impressive to permit the Istres tests to go ahead. F-BTSD had been fitted with a complete set of NZG tyres and would be put through a series of rapid accelerations, stops and ultra-sharp turns over the next several days.

As I disembarked from the aircraft that had flown me into Istres from the UK, a familiar figure was waiting at the bottom of the steps: Henri Perrier.

The 'Father of Concorde' extended his hand and we shook warmly. 'A terrible year for us to be meeting again,' he said, beaming me one of his big smiles. Not only had England won the Six Nations again, but we'd thrashed the French 48–19, sending 'Les Bleus' to second bottom in the table. They'd ended up with only a few more points than perennial wooden-spooners, Italy.

'*Au contraire*. All in all, a very auspicious year, I would say.'

'Let us hope so, *mon ami*.' Henri clapped a big hand on my shoulder and led me to dispersal, where a gleaming white F-BTSD was waiting in the Mediterranean sunshine. A huddle of engineers had gathered around one of the wheel bogies.

Four different British Concorde liveries. The manufacturer's at the top with three different British Airways ones below, spanning four decades.

Concorde in supersonic flight. At Mach 2 heat from friction and compression meant that the aircraft 'grew' in length whilst flying.

. . . and it shrank again by the time we landed. As a tribute to them, a flight engineer's cap was put in the flight deck heat expansion gap at Mach 2 on several of the last flights. On the ground, it has closed up, trapping the headwear until Concorde next flies supersonically.

Beautiful from any angle.

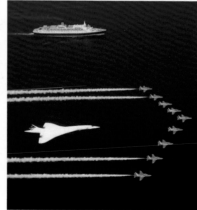

A picture beamed around the world. Air France Concorde F-BTSC struggling into the air on 25th July 2000. No other aircraft would have even got airborne.

A crash investigation includes reassembling debris to aid in the search for all important clues. As part of the British team, I clearly remember the dreadful sight and the smell in the hangar.

The 'Wear strip' that scalped the tyre of the port, inner, forward wheel in a manner not previously known in civil aviation.

My dear friend Henri Perrier checking out the new Michelin 'Near Zero Growth' (NZG) tyres.

Fitting the tank liners. Inside the tank itself. So little space. Our engineers did a fantastic job.

The first flight of the first aircraft to be modified, G-BOAF. It poured down when we arrived at Brize Norton. I made sure it wasn't called a test flight – and that I had Chris's umbrella.

Back in service at last. 7th November 2001 and G-BOAE meets F-BTSD in New York at 8:20 EST. Later that day the British Prime Minister flew on Concorde to Washington to meet the US President.

Over London on 4th June 2002 for Her Majesty's Golden Jubilee flypast. An unforgettable and moving experience. Few noticed that I was just slightly right of the centreline of the Mall. We said that it was to give Her Majesty a better view!

Running in for the Golden Jubilee flypast with four Red Arrows right alongside. I get nervous being that close to trucks on the motorway, but these are real professionals.

In my office doing 1,350 mph at 58,000 feet. From here you could see a quarter of a million square miles of the Earth's surface. Some view!

The curve of the Earth. 62,500 feet, 11.8 miles, 1,420 mph, 2,285 kmph with North Atlantic storms below. An unmodified Air Test snap taken from my 'office chair'.

30 degrees of bank at Mach 2 – 1,350 mph and so smooth that you could stand a £1 coin on its edge whilst turning.

Landing in Barbados. Concorde and her crews enjoyed a long relationship with the island. It was the only destination where we regularly reached our full 60,000-foot ceiling whilst *en route*.

The very last landing at Heathrow. The end of an era. Crowds of thousands lined the airport fences to cheer and wave to G-BOAG. It was no less emotional in the cockpit.

After the last landing at Heathrow. If you want a flag flown then I'm your man – and so is Jonathan Napier from the co-pilot's window.

Concorde G-BOAF over Clifton suspension bridge *en route* to Filton for the last ever Concorde landing, 26th November 2003, with Les Brodie and Paul Douglas at the controls, me and Warren Hazelby behind and 106 more BA staff aboard.

Gone but never forgotten. The World's Greatest Aeroplane. It truly was my privilege, and honour, to fly her.

Me in the co-pilot's seat of the Concorde simulator at Brooklands, having just guided a member of the public to fly her from take-off to landing.

Power to the aircraft before engines-start was provided by a ground power unit.

Henri and I moved to a Portakabin on the edge of the dispersal area – the team's project office for the duration of the trials.

From here we were able to get a grandstand view of the airfield as F-BTSD taxied out to the threshold, disappearing into the late morning haze.

Henri grabbed a couple of chairs from the Portakabin and set them in the sunshine against the wall facing the runway. Somebody handed us cups of coffee and me a moth-eaten straw hat to keep the sun out of my eyes. Henri wore a hat only marginally more appropriate to the occasion. Other engineers laughed: *Le Père du Concorde et son ami, le pilote anglais,* in the Provençal sunshine, as comms shot between the Portakabin and F-BTSD.

While we listened to the final checks, a squat, yellow twin-engine CL 215 water-bomber trundled by – another type that shared Istres's tarmac.

Suddenly, there was a far-off roar. I got to my feet and saw a distant curl of Olympus smoke, followed by the white outline of our shimmering bird as she appeared through the haze. F-BTSD grew and grew until, just past our position, maybe a hundred yards away, the pilot slammed on the brakes.

To a hideous shriek of metal and stressed rubber, accompanied by a cloud of blue smoke from all three sets of wheels, F-BTSD shuddered to a near-complete stop, enough to have thrown any remaining gins and tonics all over the cabin had this been a flight for real. Then, the brakes were released, and the aircraft rolled forward, slowly this time, before turning onto a taxiway.

Back at dispersal, engineers gathered around the wheels, prodding the tyres cautiously, then with increasing vigour as it became clear they had held up well.

During the afternoon, the tyres were put through further tests, including high acceleration stop-starts and sharp turns.

At the end of the day, and on the two subsequent days, the team would retire to a hotel several miles away where, over drinks and dinner, excited by the day's events, we'd talk about the tyres and how well they had performed.

On the second evening, somebody presented Henri and me with a photo, taken with a Polaroid, of the two of us leaning back in our chairs against the Portakabin, as F-BTSD had thundered by on one of the trials.

Henri laughed. Under the brims of our hats and clutching our coffees, we looked like a couple of old tobacco farmers at the turn of the nineteenth and twentieth centuries, he said.

'All we need are some *Gauloises*, a bowl of snails and a *carafe* of an unspeakable local wine and this picture would be perfect,' the veteran engineer announced to much laughter.

On the final evening, as we enjoyed the last of our red wine under the stars, his mood turned sombre, as it had when we'd met at the DGAC after Gonesse.

'What do you think will happen?' he asked, nodding in the direction of the flight test centre a few miles distant.

'I think she's going to be OK,' I said.

'*Vraiment?*'

'Yes. I have faith the tank-liners are going to work. After the success of the other parts of the mod, the liners are the last part of the safety jigsaw.'

Henri nodded. 'And the mood in Britain – at British Airways?'

'Mixed,' I admitted. 'I have a ten-pound bet with one of my captains, who says she's not going to fly again. But after the last few days, I'm confident I'll keep my money and take his.'

'I admire your *bravado*,' he said.

'What's troubling you, Henri?'

'What is troubling me is the mood in France. This aircraft is something of which every person in this country can be justly proud. In spite of the terrible tragedy at Gonesse, she is fundamentally safe – you and I both know that.' He paused, 'You are right, Mike. The fixes are good and I think that she will fly again, as well. But I think also that this will not be the end of her story.'

'The report, you mean?'

He looked at me. 'The report, I am sure, will not contain the story of what really happened on the 25th of July last year.'

I then told him about our suspicions that the tanks had been over-filled and asked if he was aware that such a thing had been going on at Air France.

'I have heard rumours, but that is all.'

'If we could prove that Air France had been over-filling, that would exonerate the aircraft.'

'Trust me, Mike, the *préfet's* focus will be on that piece of metal.'

'And then?'

'The *préfet* needs his trial. It is crazy, but this is how it is in France – the system needs to find something to blame. Some-one . . .' His voice trailed away.

'Continental, you mean?'

'The Americans, for sure,' he said, directing his gaze toward the ground. 'But it won't end with them. Afterwards, they will come for the aeroplane herself. And that means, *mon cher*, that they will also come for me.'

31

At BA, we had planned initially to bring Concorde back into service in October to coincide with the turn of the season.

The programme of mods on *Alpha Foxtrot* had come together extremely well, but before we could re-introduce a Concorde *service*, we had to do two things: first, the mods would need to be tested on the aeroplane in-flight – and then approved by the CAA – so that it could regain its Certificate of Airworthiness; and, second, BA would need to reactivate all the parts of the business that had previously supported Concorde to ensure we could still adequately operate her in airline service.

The first of these milestones would have to be achieved through an actual test flight. The second, if we got that far – and we were confident we would – would be to run what we were terming an 'operational assessment flight', a title I'd come up with to avoid any suggestion to Concorde's customers that the aircraft had to be 'tested' – with all the connotations that came with that word – in airline service.

Throughout the year, we'd tested the integrity of the three primary areas of modification – the tyres, the power supply to the wiring and the tank liners – but the time had now come to trial them in the air, with a special focus on the Kevlar liners.

Only then, if everything went to plan, would we be able to get that vital piece of paper – the Certificate of Airworthiness – that would get the aircraft back in service.

One of the tests we had to do on the fuel side was to run a tank dry to make sure, with the new liners in place, we could still get all the fuel we needed out of it. There would always be a small residual amount left – what we called 'puddling' in places – and running a tank 'dry' never actually meant there was nothing in it. But the operating manual said we could expect 99.6 per cent of what we'd put in to come out, and we needed to check that, with the Kevlar linings in place, that was still the case. The only way to do this effectively was to run an engine until it shut down through fuel starvation – and under 'dynamic' rather than 'static' conditions, which meant 'in the air'.

This, in turn, meant the aircraft would need to land on three engines; and because we always had to plan ahead, if we were counting on landing on three, we had to contingency-plan on the basis we might lose another one for real.

This wasn't a big deal from a technical standpoint, but deliberately planning to land at Heathrow on anything less than a full complement of Olympuses was a no-no, which meant we were going to have to look to land elsewhere.

We had got to the point where the accident investigation and the return-to-service programme had converged. As far as we could tell, the impact of the mods had been far less than we'd expected. The increased weight of the Kevlar linings had been beautifully and fortuitously offset by the decreased weight that had accrued via the installation of a new lighter weight British Airways cabin interior.

Even an additional 100 kg would have had quite a detrimental effect on revenue. A hundred kilos equated to a single passenger and their baggage. Take that away and we were talking about a $10,000 loss of revenue on a round-trip. Multiply that loss over a year of operations, and it would be considerable. Around $6,000,000. The new interior more than compensated for this, however.

It was now about ticking boxes. Engineering had already made a pitch to the BA board that only five of the seven aircraft fleet should receive the mods – this, they said, was the maximum number they foresaw the scheduled operation needing – and it was. What this did was sound a death-knell for the charter programme. At that point, I wasn't overly concerned because I figured that down the line, after we'd rebuilt and stabilised the scheduled operation, we could look at reinstating the charter business. Sadly, however, it never got to that point.

The two CAA inspectors who'd been assigned to the fleet were Jock Reid and Gwyn Williams. Both of them were highly accomplished pilots – both had previous considerable experience of Concorde. Gwyn was our operations inspector for Concorde. Jock was the CAA's Chief Test Pilot. Jock was tall, grey-haired, a bit older than me, utterly pragmatic and the antithesis of the dour Scot. I liked him immensely, recognising early on that, contrary to some regulators, he was the kind who always tried to say yes rather than no. You could tap him and Gwyn for advice and support in any area and know that neither saw it as an admission of weakness. It was, in my opinion, how all our regulator relationships should have been – sadly, not all of them were.

As part of their regular CAA inspectors' role, both Jock and Gwyn had flown Concorde on the route – i.e. with passengers – from the co-pilot's seat. On the route, as CAA pilots, they were confined to the right-hand seat as co-pilots and always with a training captain in the left-hand seat. Outside of on-the-route flying, Jock, in particular, was always looking for left-hand seat experience and I reasoned, with the test flight coming up, this was one of those occasions. Whilst he would be my co-pilot for the flight, I decided that he should sit in the left-hand seat and fly most of it himself.

For the flight itself, the CAA had granted me provisional test pilot status as the aircraft commander.

As part of my job, I was BA's Concorde Chief Test Pilot, but that was for the aeroplane's regular Certificate of Airworthiness programme. Each of our seven Concordes had needed a major check every five years, at the end of which we had to take them up and put them through a very thorough evaluation programme to make sure that everything worked as it was meant to, even outside the normal operating envelope. For *this* flight, I needed a special dispensation from the CAA because, unlike all the other times, we'd be putting the aircraft through regimes and manoeuvres that were still further outside its normal operating envelope. This included deliberately running one of the engines dry, plus trialling the other systems affected by the mods.

A date for the flight was fixed: 17th July. We could fly out of Heathrow. All we still had to decide was where we located our support infrastructure – and, *de facto*, where we landed, these two being one and the same thing.

Was I nervous? Despite having mitigated as many of the risks as we could, there were still things that could go wrong. This would be the first time that anybody had flown a BA Concorde outside of the simulator for almost a year.

Second, the engineering and test-flying requirements were out of the normal for us. We'd spent a lot of time focused on these aspects, less on how we would get the aeroplane back into normal operation – a bridge we'd cross later.

Furthermore, if anything did go wrong, I'd need to

distance myself from my emotions and report everything – the good and the bad – exactly as I saw them.

It would be hard to be objective about an aircraft that I wanted to see back in service – but that was what the job demanded of me.

Lastly, all of this would be done in the glare of the media. Being the first flight of a modified Concorde since the accident, we knew it would generate a huge amount of coverage. We, though, had to strike the right note throughout. Yes, we had to mark the fact that Concorde was flying again, but we could never forget either that the flight and the mods that underpinned it were down to the fact 113 people had died.

The PR people recognised, too, that this wasn't an easy path to tread. We also had to get the message over that the aircraft had been through a major modification programme without insinuating – because it wouldn't have been true – that the aircraft had been unsafe to begin with. But nor did we want to denigrate our French brothers and sisters, the only other Concorde operator, upon whom the tragedy of the accident had impacted directly. At the heart of it, we needed to hold them and the 113 uppermost.

In the end, of course, the location was a no-brainer.

Somewhere that had a long runway; that wasn't too far from Heathrow; that could accommodate our large engineering team; where we could assemble the integrated product team (including reps from the manufacturers, engine producers and systems providers), regulatory authorities and accident investigators, which also had its own strong engineering infrastructure.

If it felt like we'd been here before, it was because we had: we were going back to Brize Norton.

32

Over the next several weeks, keeping my air-accident investigator's hat on in the build-up to the flight at Brize, I worked with my British colleagues to try to decode the crew of AF4590's fuel actions as they'd prepared to depart CDG.

A tank had a sensor-switch that indicated when the fuel level had reached 95 per cent.

Under a normal refuelling operation, the refuelling computers would close the valves as soon as the sensor signalled the level had been reached. They were designed specifically to stop you filling to more than 95 per cent. The valves were operated semi-automatically by the refuelling system and also semi-automatically by the system we used to transfer fuel in flight.

But like most things on an aircraft, there was a manual back-up – a switch that allowed a human-in-loop to override the valves in case the automatic system failed.

If it did, another sensor flashed up a warning at 97 per cent. If the 95 per cent cut-off failed for any reason, the tank would carry on filling to 97 per cent, at which point a warning light would come on. This said: 'O/FULL' for over-full. You'd then know to take action: you had to manually shut the valve to stop more fuel going in.

Here, though, was where things got interesting.

If you could manually close it, you could manually open it, too.

The unlicensed procedure I'd informed my fellow

investigators about, which had been used by some crews in Concorde's early days at BA, was to pump fuel forward, then manually open the valves and allow fuel into certain tanks above the 95 per cent level. Then, when the over-full light came on, to have closed the valve manually at 97 per cent. If you were tight on fuel, there was a considerable advantage to this because you didn't have to burn off this additional fuel before take-off. OK, you were up to 97 per cent in some of your critical tanks, but that was all right because you still had that all-important air gap.

It was a procedure that wasn't hard to carry out either – what you had to do was switch a certain combination of pumps and valves, and job done.

But what it required was for the flight engineer to monitor the whole process very closely, because now the only thing stopping the fuel going above 97 per cent were his eyes on the panel. When that 97 per cent O/FULL light came on, it was down to the engineer to shut down the valves and pumps. If he missed it, the fuel would carry right on pumping forward, because the automatic cut-off system – set to kick in at the 95 per cent level – had already been bypassed.

As a young Concorde co-pilot, I'd been aware this had been possible – and, more than that, that it had occasionally been used as an unapproved, not-written-down-in-the-manual way of getting more fuel aboard at take-off and/or saving time otherwise wasted having to burn the valuable stuff off.

When I came back to the fleet as the Chief Pilot, I was aware that it was still occasionally going on – and that some of our training engineers were demonstrating it to our flight engineers as something they might consider for fuel-critical sectors. As pilots, we continued to be aware of it, too; and, on one level, it made sense: extra fuel on certain marginal legs could be very useful.

So, one of the things I did when I became the boss was to say, 'OK, I know about this unapproved procedure and I'm uncomfortable about it, so here's what we're going to do: we either get it approved or we stop doing it. I don't mind which, though I'd actually prefer to get it approved because it has operational benefits. But we do one or the other – we can't carry on doing it as we are.'

So, we went to the manufacturers and described the procedure to them. They then took the data away and promised to come back with their answer. It didn't take them long. They declined to approve it for a very simple reason: success, they said, was solely dependent on the flight engineer monitoring his panel and closing the valves and pumps when that light lit at 97 per cent.

The light itself was hooked up to a single sensor and went to a caption-reader with two bulbs in it. The chances, thus, of a double bulb failure were slim – we checked the bulbs every flight.

The chance of the 97 per cent sensor failing was greater because it was a single sensor – its sole purpose was to indicate over-fill, which shouldn't have happened anyway because the tank was designed not to be filled to more than 95 per cent.

But the biggest weakness in the system, the manufacturers told us, was the fact that it relied on a human being to monitor it and switch it off when the 97 per cent light came on. If the human-in-the-loop got distracted and didn't notice, then the system would just carry on pumping fuel forward.

And it wouldn't just carry on until the tanks were full. When they hit 100 per cent, the fuel would be vented out into what was called the 'refuel gallery' – which also served as the 'fuel transfer gallery' and/or the 'fuel jettison gallery'.

The gallery ran the length of the aeroplane and was designed for moving fuel around the aircraft and to mop up any that over-spilled during refuelling or the fuel-transfer process.

The designers had pretty much thought of everything. There was a 'scavenge system' in the gallery that syphoned the collected fuel into a 'scavenge tank'. The tank was small – it held around three gallons only; a scavenge pump automatically removed any fuel that had collected in it and returned it to Tank 3. After the scavenge tank, there was nowhere else left for the fuel to go – it would have to have been vented out via ports in the rear fuselage.

The point was, I told my colleagues, you could, in theory, fill the tanks to 100 per cent this way.

Furthermore, I said, the tanks would then be pressurised, because the pumps would try to put fuel into the tank faster than it could physically overflow. Plus, if Air France had been over-filling them for any length of time – conceivably, even, for the entire length of time the aircraft had been in service – twenty-four years – then it was possible the metal of the tank had become stressed.

The question we then asked each other was: had there been evidence at the crash site of over-filling? Could over-filling have occurred on Flight 4590?

This was the critical question. In order to take the investigation to the next stage, like a police profiler trying to establish the motive for a crime, I had to establish the mindset of a crew that would have been capable of missing an O/FULL light.

33

There was plenty of evidence pointing to the fact that the crew were under pressure and had been in a rush to get away.

Flight 4590 was a charter flight carrying a hundred passengers from Paris to New York as the summer season moved into its height. The temperature had been 19 degrees Celsius; the surface wind light. The passengers were going to join up with a cruise ship in New York, so the baggage load was high too; most of them were planning on being on the cruise for around a fortnight.

The Concorde that had been originally allocated for the flight had had a technical issue overnight, and so it was changed for another one. In turn, that one had developed a problem with one of the thrust-reversers, which wouldn't deploy properly. Given this, as the captain, you had a three-fold choice: swap the aeroplane for another one – but that had already happened and there wasn't a third Concorde available in time. Or you accepted the reverser wouldn't work properly and locked it into position for the flight (the thrust-reversers, it may be recalled, also modulated in flight to assist in the balancing of the two turbine shafts within the engine). This wouldn't affect the aircraft's handling or take-off performance too much, but it would have quite an impact on range – and the associated penalty would not have been insignificant.

The third option was to wait for it to be mended, which clearly took time.

The first option simply wasn't possible.

The second option wouldn't have worked either because this was a fuel-limited sector – the plane was full and every last drop of fuel on board needed to be available. An adjunct to this option was that you accepted the locked thrust-reverser scenario, and accepted the fuel penalty, with the proviso you made a technical stop *en route* to refuel. But, then again, a technical stop would have been ruled out on time grounds – the minimum amount of time it would have taken at Shannon or Gander (by the time you've slowed down, landed, refuelled, taken off and sped up again) would have again caused the passengers to have missed the sailing of their ship.

So, the captain, rightly, decided to ask for the aircraft to be fixed by changing the motor that drove the thrust reverser. We knew this from records held by Air France maintenance. Getting the fix done took around two hours. The two-hour delay heaped additional pressure on everyone to get Flight 4590 away as quickly as possible. The pressure was significantly greater on this flight – a charter flight – than it would have been had AF4590 been a regular scheduled operation. On a conventional flight, we always tried our damnedest to avoid making it late, but it wasn't critical if we did – by and large, people were no more than slightly inconvenienced. On a charter flight like this, though, it was different – if the flight was late, everyone missed their ship, effectively their hotel – and their holiday cruising to South America.

Everyone involved in the flight would have been feeling that pressure – the ground-loaders, the refuellers, the crew . . . everybody. And there was evidence of mistakes having been made across various areas of the dispatch.

For instance, big bundles of newspapers had been put on

board – sixty kilos or more – that were never entered onto the load sheet. The last suitcases were piled on – nineteen in all – without having been entered on the load sheet either. We knew this because, even though all the load sheets had magically vanished, the investigators had constructed their own by reverse-engineering the data: talking to ground-handlers and working backwards from data that *had* been retained to put together a working substitute. From this, they'd discovered the papers and extra cases thrown on at the last minute.

Both of these omissions were more serious than they appeared.

Concorde had two holds. The lower baggage compartment, the smaller of the two, was accessible via a belly-hatch just aft of the nosewheel. The upper baggage hold – the main baggage compartment – was located at the back of the aircraft, beneath the fin, just aft of the passenger cabin and Galley 7 and a long way behind the aircraft's centre of gravity. Because it affected the C of G less, the bags were put in the lower hold first. And because it was much smaller than the other hold, the lower hold would have filled quickly – and on this flight, which was at capacity because people take a lot when they're going on a long cruise, it would have rapidly led to bags being loaded into the upper.

Being at the back, the upper hold was critical to the C of G. This, then, made it all the more perplexing that those nineteen bags had not been recorded on the load sheet, as investigators had discovered from talking to those involved, because they would have had a marked effect on the C of G, pushing it further aft than anyone thought. In my discussions with other investigators, I'd previously asked them to consider a nominal shift to 54.4 per cent from the standard position: 54 per cent. With the additional, unregistered weight of those bags, I now asked them to consider the possibility that the all-important C of G had shifted by more than a whole percentage point – to 55 per cent.

Computers are only as good as the information they receive. The computer would have known all about the aircraft in question, tail number F-BTSC – *Sierra Charlie* – starting with its empty weight, the fixed equipment on board, where the galley equipment was *et cetera*, and, from this, its zero-fuel C of G position. All these data-points would have been different from the other Concordes in the Air France fleet – there were differences across the *whole* fleet – and these differences would have been recorded by the computer. As would the variables on the flight: the crew, where they sat – right down to their specific seats.

The system knew, too, that the aircraft's on-board computer would precisely monitor the refuelling operation and the fact that particular tanks would be filled in a carefully choreographed sequence.

It then needed to know the number of passengers and where they were seated within the cabin's four zones, designated A, B, C and D on the sheet.

Next, it needed to know about the baggage – the number of bags in the lower and upper hold and the weight in each.

Once all that had been logged, it could then calculate the C of G.

But if you neglected to inform the computer that there were nineteen additional bags in the aft compartment – as had happened on AF4590 – the computer would be just as ignorant as you were, causing it to assign a false C of G position, an error that would have been compounded, too, by the 60 kilos of newspapers that had been chucked on board, also without being formally recorded. There had been nothing malign or intentional about this; it had just happened, as human error tended to. But its impact that day had been crucial.

And then there was the matter of the fuel.

All commercial flights were refuelled with the amount they needed for the flight, with some extra for holding and diversion at the destination, and some for 'contingencies' – the unknown and the unexpected. Altogether those elements made up 'flight plan fuel'. Airlines differed slightly in how they assigned flight plan fuel; we were now having to get to grips with how Air France assigned its flight plan fuel.

On top of flight plan fuel, you also – as by now even the non-pilots amongst the investigators were intimately aware – needed taxying fuel.

This amount wasn't fixed. It was up to the captain at the briefing stage, pre-flight, to say how much was required and the amount would vary depending on where you were and the circumstances of the flight.

If you were at JFK, you were delayed, and you knew it would take forty minutes to taxy to the threshold, you might want as much as 4 tonnes extra.*

If you were in Barbados, on the other hand, where you might be the only plane in the queue – where you simply taxied out and took off – you'd maybe keep it to a tonne. Of course, you could only choose the higher amounts if you had the space available.

The point was, the fuel you put on board for taxying didn't *have* to be burnt off pre-take-off. You planned your take-off

* Taxy fuel is burnt at 6 tonnes per hour, so forty minutes is 4 tonnes.

weight, you put taxy fuel on top, but within the regulations it was OK *not* to burn it all off – as long as you'd got the space for it, you could take off with almost all the taxy fuel you'd not burnt.

We knew that the crew of AF4590 had put 2 tonnes of taxy fuel on board, which wasn't unreasonable. From their point of view, it was in their interest to minimise taxy time in order to burn off as little fuel as possible as they went out to the runway.

From the flight data recorder, we could see that, of the 2 tonnes extra on board, they'd only used 800 kg of it – this also from the cockpit voice recorder, from which Jardinaud, the engineer, just pre-take-off had answered Marty's question about how much taxy fuel they'd burnt – giving them an additional 1,200 kg of fuel that wasn't accounted for on the load sheet. Handy additional fuel to have aboard on take-off if you needed every drop you could get.

To hang on to as much as possible, it looked as if they had cut corners – from the cockpit voice recorder, we could hear that they'd done some things on the taxy-out that weren't done in the way they should have been, perhaps the most serious of which had related to a failure in the fly-by-wire system.

Concorde had two FBW channels – Blue and Green. You normally operated in Blue, with Green as standby. They were otherwise identical.

If a problem occurred with Blue, it switched automatically to Green. That would bring on an associated red light and sound a warning gong – procedures said you should then refer to a checklist.

In AF4590's case, the problem was with the rudder set – Concorde had three sets of control surfaces: (1) the inner and (2) outer and middle elevons on the wings and (3) the rudder – which had registered a fault. The result was that the FBW system had automatically switched. The crew reset to Blue and started a discussion about what to do but it was curtailed. Shortly after that the rudder switched to Green again. Another discussion – again curtailed.

No checklist had been called for. Good, conscientious airmanship said you should at the very least check the 'Minimum Dispatch List' and/or 'Dispatch Deviation Procedures' to see what advice came with this part of the system misbehaving prior to take-off.

Had I been doing a sim check, not looking at the checklist would have been a 'comment point'; not referring to the Minimum Dispatch List would have been a 'discussion point' – the latter is a bit of advice; the former signified a mistake.

Marty and his crew did neither of these things, although what they did do turned out to be correct – they 'hard-selected' the affected channel to Green to back up the automatic switch-over. Technically, that wasn't surprising, because the co-pilot, Marcot, was an instructor and it was virtually certain he would have known the procedure – naturally, without recourse to the checklist. But that wasn't the point. The point was it should have been checked via the checklist and it should have been discussed amongst them.

Had I been carrying out a sim check, I would have marked the fact that it wasn't discussed as a possible 'whole crew issue' – evidence, perhaps, of some generic indiscipline caused by the take-off that day being rushed.

But, so rushed that they had implemented the unauthorised fuel transfer procedure and for the flight engineer to have then missed the O/ FULL light?

To have made this deduction with any degree of confidence, I needed some evidence from the wreckage that pointed to it – and, as I had witnessed with my own eyes, there was precious little of the cockpit that had survived.

But of several pieces that *had*, two had shown some curious anomalies.

One of them was a so-called 'lock-toggle switch' from the flight engineer's fuel panel that had been found in the 'up' or open position at the crash site – not what you'd expect if the fuel-transfer procedure had been done by the book.

The second was a gauge, frozen into position upon impact, that had pointed to there being considerably more fuel in Tank 5 than you'd have expected given the minute-and-a-half duration of AF4590's tragic final flight.

34

The next day, I found myself in the simulator at Filton.

The original sim, built and installed on the site of the BAC Concorde final assembly plant in the mid-seventies, had been upgraded in 1987 with an all-new visuals suite – one that had replaced the clunky (but then-state-of-the-art) platform in which a camera had followed the pilot's inputs by roaming over a series of 1/2,000 scale model landscapes, including an airport that was a lot like Heathrow and a cloud-top model for our high-altitude supersonic work.

Most of the cost of the upgrade had gone on a full daylight panoramic visuals system that gave pilots a 150-degree by 40-degree view of computer-generated images projected onto a TV display in front of the windscreen. Thanks to the virtual nature of the environments, the new visuals allowed us to practise take-offs and landings at most of the airports we were ever expected to operate to and from – a vast improvement over the original.

Sitting 15 feet off the ground on six hydraulic jacks, the sim was an intimidating-looking beast, but once inside it looked and felt familiar and cosy – exactly like the real thing.

After you'd gained access via a retractable gangway, you settled into the sim detail, with your instructors in the jumpseat or monitoring remotely.

After this particular session, in which I'd gone over key portions of the test that we'd soon be undertaking for real, we broke for the day. A final chat with the crew and then I'd be heading home – a ninety-minute trip back down the M4. I heard the flight engineer's footsteps on the gangplank, then he poked his head around the door. 'Kettle's on, Mike . . .'

'Be with you in a moment,' I said over my shoulder.

Alone again, I got up and moved to the engineer's seat.

I looked at the panel.

Spread across it were a mosaic of switches and dials. In the middle was a section dedicated to the movement of fuel. At the centre top left, in amongst a bewildering array of pump selector switches and their associated gauges, was a panel dedicated to the standby inlet valves for Tanks 5, 6, 1 and 2. The switch for Tank 5 was on the left.

Tentatively, I moved my hand to it. I hadn't flown the aircraft for a while, and I wasn't a flight engineer – this part of the cockpit was peripheral to my knowledge. I needed to convince myself before I tried to convince any of my colleagues that my theory held up to scrutiny. And it hinged on this switch.

On a conventional switch, a light switch, say, you had two movements – on and off. Say, up for 'on', down for 'off'.

This switch had three positions: a central, neutral 'off' position; an up position for 'open'; and a down position for 'closed'.

The inlet valves were such crucial switches that the designers had built safety features into them to ensure that they couldn't be activated by accident.

To open the standby inlet valve for Tank 5 – to allow it to accept fuel under pressure from Tank 11 – you had to do three things beginning with a clear, conscious thought: *I want to open this valve.* You couldn't just flick the switch – you had to first pull it towards you, which I did now – and because it was spring-loaded, I had to pull it hard. This required a lot more force than I had imagined, but with it now 'out', I performed the next action: I moved it up. Then I released it. The switch snapped back into position and stayed there, locked in position.

The valve for Tank Number 5 was now 'open'.

Lock-toggle switches were known for their resistance to movement under the most violent of impact conditions. To be sure of this one, I tried to pull it down now. I put a lot of force into the action, but it wouldn't budge.

I left the sim and joined the crew for the tea and biscuits they'd prepared. We talked about the session, but a part of me was still in the sim.

Had Jardinaud forgotten to close the valve to Tank Number 5?

The evidence from the wreckage was clear – it had been found in the up/open position. It had gone past 97 per cent. *Had he missed the O/FULL light?*

Had he closed the valve, the evidence from the crash site would have shown it.

Why was this significant?

Tank 5 was the tank that had ruptured after the wing had been struck by the piece of rubber. It was a main transfer tank feeding Engines 1 and 2.

With fuel pouring out of the A4-sized hole in the underside of the left wing and the two Olympuses guzzling what *had* been in it at a phenomenal rate, you'd have expected Tank 5 to have been virtually empty at the moment it crashed. But the gauge showed it had been almost 30 per cent full. How could it possibly have had so much fuel left in it?

Only if the pumps had still been pumping fuel forward from Tank 11.

Driving back down the M4 for home, I ran through it, over and over.

There were two elements to this: first, determined to carry

all the fuel they could, the crew had decided to initiate an unapproved fuel procedure; second, they hadn't shut the procedure down. Those pumps had continued to pump as Flight 4590 had rolled down the runway; and they'd carried on after it had hit the wear-strip.

With fuel gushing from the tank, and the crew battling to save an aircraft now on fire, all their attention had gone on trying to prevent the unfolding disaster.

And despite fuel still pumping into Tank 5, it was losing fuel overall. As it was forward of the C of G, this loss made AF4590 increasingly unstable.

There was, not unnaturally, a pervading desire on the part of the French-led investigative team not to blame the crew – something we understood and sympathised with. No one wanted to blame anybody who couldn't be there to answer for their actions.

Over the coming days, we tried to persuade our French counterparts that this new evidence deserved to be taken seriously in the spirit of all air accident investigations – because we needed to get to the truth. But from the mood music in Paris, it became increasingly clear that it wouldn't be heard.

The BEA was sticking to its mantra: that the tank had ruptured from within because it had only been *effectively* full.

By this, it wanted the world to believe that the Number 5 Tank had been filled correctly, but because of the aircraft's acceleration down the runway, the fuel had surged to the back where the rubber had struck the underside of the wing – and that this had set up the conditions for the disaster: the fuel having nowhere to go except out through the tank wall. A fudge of classic proportions that might have been funny were it not for the terrible tragedy that had taken place.

35

With intransigence and denial rampant in Paris, we did the only thing we could: we rechannelled our frustration into the mod programme at BA that would see Concorde take to the air again for the first time in the UK.

The crew on the day were Jock Reid, me, our flight engineers Robert Woodcock and Trevor Norcott, with Les Brodie in the jump seat. Behind us, were a team of engineers from BA, the manufacturers and their suppliers. Jock, as I'd promised him, would sit in the left-hand seat for the flight and I'd sit in the right.

I was still the aircraft commander, though, so I had to sign the 'Technical Log' to formally accept the aircraft for the flight. It was one of the proudest moments of my career. I'd do the take-off from Heathrow and then we'd head out into the Atlantic for the very extensive and detailed schedule of multiple tests. The culmination of the flight would be an engine-out landing at Brize, having demonstrated along the way, we hoped, that the new liners in the fuel tanks didn't degrade the engine-fuel performance or the aircraft's.

Unusually, I got a weather briefing on the morning of the test before I left home. Normally before an en-route flight I'd check the weather forecast on the TV the night before we flew, then again on the radio in the morning. Then, upon arriving at Heathrow's Briefing Centre, I'd get another update, this time with all the bells and whistles: weather conditions on take-off and landing, on-route, wind direction and strength on the flight and so on. Today, however, I got a full briefing on the flight we were about to undertake before I left home, with a particular focus on the state of the mods and the serviceability status of *Alpha Foxtrot*. The weather at Heathrow was glorious, but the weather that was predicted for Brize when we landed was awful – wind and heavy rain, lots of it. This was OK for the test, but something to hold in the back of our minds. The people I felt sorry for were the PR bods and the media who'd be waiting for our return in the stuff. I suddenly remembered that that would include me.

When I mentioned this to Chris, she came up with a top-notch idea.

'What you're going to need is a great big golfing umbrella,' she said, and disappeared for a few moments. When she reappeared, beaming a triumphant smile, she was holding two big Concorde umbrellas we'd been given as part of some promotional gig a few years earlier. 'When you're walking across the tarmac and it's hoofing it down and it's all on telly, you'll not only be bone dry, but sporting the corporate logo. Job done!' She kissed me as I turned for the door and said: 'Oh, and Mike, I like them. Don't lose them, will you?'

I promised her I wouldn't.

A couple of hours later, I lifted Concorde into the air from Heathrow and banked her towards the west. Whilst I never forgot why we were doing this, the intervening months since the accident momentarily slipped away and a part of me had to resist setting course for New York or Washington.

When we were out over the Atlantic, we swung to the right and routed up towards Iceland. We would get to a point 300 miles off the coast before swinging round through 180 degrees and heading back to the UK. All at twice the speed of sound.

The idea was to simulate a New York sector by being supersonic for the vast majority of the three-hour twenty-minute flight. But, operating on a CAA flight test certification, we weren't authorised to enter Canadian or US airspace. Hence the routing, which kept us in an area governed by joint UK and Irish control.

I'd love to be able to report – harking back to George Edwards and the woman who'd felt short-changed by her Concorde experience flight – that our test-flight in *Alpha Fox* had been filled with moments of high drama and tension, but the truth was it went like clockwork.

After three hours, Jock brought her into a wet, windy and squally RAF Brize Norton for a near-perfect touchdown on three engines. The aircraft had outperformed all the expectations that we'd had of her. We taxied in, shut the aircraft down and filled in the extensive paperwork that needed to be completed. Then it was time for me to go and debrief the press.

Two of the first people on board to greet us were BA's Commercial Director, Martin George, and the station commander of Brize, a tall, slim, dapper group captain. They briefed me on the plan, which was for the three of us to walk over to the main terminal in the downpour, where the media, drenched from the rain, were gathered. Martin and the group captain were holding black, business-style umbrellas, but from the look of the rain that was lashing the windshield, they didn't seem up to the job.

I grabbed the two Chris had given me and we proceeded down the steps.

On the long walk to the terminal, the group captain, who towered over Martin and me, suggested that he hold my umbrella. This made sense as I was going to be the one doing the talking and the group captain, being taller than both of us, had enough elevation to keep both Martin and me somewhat protected. What a nice chap to think of that, I thought.

Gathering myself before the reporters ranged in front of me, and raising my voice against the elements, I explained the purpose of the test and that it had been a complete success. I acknowledged the terrible accident that had led to this moment but added that this was a vital step on the road to ensuring nothing like it would happen again – and that, all being well, with the test concluded, it ought to be no more than a formality now that Concorde would get back into scheduled service. Air France had announced its intention to restart in October, but, I said, it was our aim to start a little sooner, in September.

After answering a few questions, I headed towards the terminal where it was planned that I would do a few one-on-one interviews for the evening news.

Before stepping into the dry, I turned from the huddled press for a quick look across the airfield. The view and the backdrop – a gathering of RAF transport aircraft out on the apron behind *Alpha Fox* – was redolent of the first time I'd seen Concorde fly – the day the world had watched – that day, when Raymond Baxter of the BBC had provided his clipped, yet surprisingly emotional commentary of her landing at nearby Fairford, with Trubbie at the controls, after that first, historic short hop from Filton.

And, of course, it had been here at Brize I'd had my first experience of flying her too.

I took a breath and stepped into the warmth, readying myself for my first interview.

Just then, my phone buzzed. A text from Chris, asking me

how it had gone and sending me her love. I smiled and put my phone away just as it hit me. *Bloody hell, my umbrella.*

I looked around for the group captain, but he was nowhere to be seen. I tried to convince Chris it was hers I'd lost but, strangely, she didn't buy it.

A quirky end to a great day. I got home to discover that 136 MPs had signed a 'Commons Early Day Motion' congratulating the whole team and that they were looking forward '*to the successful resumption of normal Concorde services by both British Airways and Air France.*'

Despite the bad taste in our mouths over the incomplete investigation, it was clear there had been some shift in the ether – that the tide had turned.

A couple of weeks later, I drove to the CAA's offices at Gatwick Airport. It took me a while to find the office of the issuing officer, a room, when I entered, that was little bigger than my downstairs cloakroom. After chatting for a few moments, he looked up and said: 'I'm delighted to tell you, Mike, that we're going to reinstate Concorde's Certificate of Airworthiness.' He produced a piece of paper from a drawer, signed it, stuck a stamp on it and passed it to me. I looked at it and thought: *this is fantastic, but shouldn't there be some sort of celebration to the accompaniment of trumpets for this moment?*

I shook his hand, thanked him and walked out into the summer sun.

It wasn't until I got to my car that I noticed the date on the certificate the CAA had selected to mark its validity: 16th August. Whether or not they had chosen it deliberately, it was a year to the day after it had been suspended.

Less than a month later, the time came for our own dress rehearsal.

The mod programme had been a total success. We had our Certificate of Airworthiness. Concorde was back in the air. The big question for us was could we, BA, still mount a successful Concorde service?

To answer this, we decided we needed an internal test-flight – something that went beyond the workings of the aeroplane to test the functionality of the system – from booking the tickets, checking passengers in, and testing the in-flight service – inside the airline's own loop, exercising all the things that hadn't been exercised for almost a year. To do this, we would run an actual Concorde flight for real, with a full complement of passengers on board. Only we couldn't take it to the US, because the aircraft hadn't yet been granted its US Certificate of Airworthiness, so we couldn't put the fare-paying public on board and we couldn't call it a 'test-flight' because the media would freak. So, we did the next best thing – we elected to fly passengers on a three-hour twenty-minute leg, again as if it *were* the New York run, with a hundred BA employees on board, most of them engineers. This was to be our 'operational assessment flight', that anodyne term that sounded suitably techie but, hopefully, lacked adverse implications.

We took off on a beautiful bright September morning and headed out towards the Bay of Biscay.

The mood on board was an odd mixture of upbeat and mindful – the former because this was a great way of thanking all those who'd helped get Concorde back into the air; the latter because none of us wanted to forget those who'd lost their lives on AF4590. We broke into Mach 1, accelerated through Mach 2 and cruised towards our maximum ceiling for a representative amount of time had this been a real, scheduled flight, while the passengers in the cabin enjoyed the fruits of the Concorde service we'd launched in the interim amidst the new interior that we were keen to show off as well as test.

It all went like clockwork. Back on the stand, we went through the shut-down routine on the flight-deck (one that I and other crews had been through repeatedly over the past several months as it became clear the aircraft was approaching service re-entry), as, behind us, I heard the cabin crew saying cheery goodbyes to everyone who had travelled with us – just as they would have had they been real, fare-paying customers. I turned just as the last one, an engineer called Ricky Bastin who'd been with us for many years, was being waved onto the Jetway. Meanwhile, a group of senior execs was coming in the other direction – a top-level delegation composed of the chief exec, Rod Eddington, and a few other dignitaries. They'd hardly stepped on board when we heard the sound of someone running down the Jetway behind them. I was surprised to see it was Ricky, out of breath and ashen-faced.

'Have you heard?' he said, his shock at whatever he was about to tell us eclipsing the fact that we had the chief executive and his team on board. 'A passenger jet has just crashed into one of the Twin Towers in New York. An American Airlines' 757, they're saying. The pictures are just hitting the news.'

36

In the wake of the air traffic downturn that resulted from the 9/11 terror attacks – transatlantic routes alone were down 30 per cent on their pre-terror attack levels – we decided to postpone Concorde's return to passenger service for two months. When the big day came, on 7th November, Air France and BA had agreed we'd resume operations with coincident flights to New York, emulating the simultaneous take-offs our two airlines had made when Concorde had entered service just over a quarter of a century earlier.

As with everything we did after the accident, we had to consider the tone of the service re-entry event – respectful, but not dour. It couldn't come across as downbeat, but nor could it come across as overly triumphant.

But we at BA also had seats to sell. The cost of the modifications and the revamped cabin interior had amounted to around £28 million, just to us, which we had to recoup; so, inevitably, a PR and marketing campaign would need to accompany the 7th November relaunch. As part of the preparations, I travelled to New York to meet with our marketing team and a deputy from the office of Mayor Rudy Giuliani, who had signalled that he wanted Concorde's return as evidence his city was back in business – that nothing, not even an outrage of the magnitude of 9/11, would stop the Big Apple from getting back to normal.

More than six weeks after the attacks, when I went to

Ground Zero, the dust from the two collapsed towers still hung heavily in the air.

My mind flashed back to the time Chris and I had gazed over Manhattan from the Windows on the World restaurant on the 107th floor of the North Tower.

Memories of a happier time, a different world. As I journeyed between meetings, I tried to focus on these, not the images seared into our minds from the images on the TV.

Back in London, at BA HQ, we discussed who should be on that first flight and decided that it needed to be a mix — some would be BA staff who had been integral to the modification and return-to-service programme, a way of our saying thank-you to them; there would be some senior-level BA execs; some of our regular customers would be on the flight; a few people who'd won their seats as competition prizes; a handful of media; and some celebrity Concorde regulars. In the run-up to the relaunch, we'd not only canvassed our customers about the return-to-service but had invited our most regular repeat customers to events at Heathrow and New York, in which we'd shared with them what had gone into the mods. It had been a chance, too, to show off the new cabin interior. One of the top-level businessmen had impressed upon me at the London event just how vital Concorde had become as a business tool.

'I'm not fussed what you've done on the technical side, Mike. We know the aircraft and have absolute faith in her — and in you guys too. What I need to know is when she'll be flying again, because, without her, my company is losing money.'

We were mindful, too, that a great many of our regular customers had been killed in the 9/11 attacks.

One company, Cantor Fitzgerald, lost more than 650

employees, many of whom had flown with us. We would remember them, too, on the day we resumed service.

On the crisp November morning when that day finally came, we took off a little before a quarter to eleven in one of the three modified aircraft: *Alpha Echo*.

Sitting in the left-hand seat, I eased her towards the Bristol Channel where we would break the sound barrier on the way to our Mach 2 cruise speed and 58,000 feet cruise altitude on our great circle track to New York.

At almost exactly the same moment, an Air France Concorde had taken off from Charles de Gaulle.

Touching down at New York at 2 p.m. local time, having experienced a frustrating air traffic delay at Heathrow, we taxied to a part of the JFK complex where the Air France Concorde was already waiting, the two aircraft ending up nose-to-nose for a media photocall. Just as the passengers were set to disembark, the door flew open, and Rudy Giuliani strode into the cabin. This was as much a surprise to me as it was to everyone else. The Mayor of New York made an impromptu speech to everyone on board about what this and the Air France flight meant to him and to New Yorkers in general. They – *we* – he said, had survived the attacks and had demonstrated their resilience by signalling that the city was open to business – as it always had been, as it always would be – and Concorde's return was emblematic of the moment.

We had had a fair indication in advance that our regular customers would support us – that they had confidence in what we were doing – and we were proved right. Our loads were good – not quite as strong as they'd been prior to the crash, but that was because the market was depressed post-9/11.

But in terms of the percentage of transatlantic premium passengers travelling on Concorde versus the total number of transatlantic premium passengers BA was carrying per day, the percentage actually went up.

In the period that followed – late 2001, early 2002 – whilst getting back into service had been the real milestone, there were some difficult commercial decisions still to be made.

As we focused on making the return operation robust, ensuring the five aircraft we'd by now elected to modify (out of the total fleet of seven) would allow us to take a view as to whether we'd be able to go back to a full double-daily operation to New York, as well as some of the other things we'd done beforehand – the charter/special flights programme, for instance, and the schedule to Barbados – I'd not fully appreciated just what we were up against.

To me, it had been self-evident that we'd strive to get back to a normal operation – up to and including the airline's full complement of seven aircraft.

But cost/benefit analyses going on in the background were showing the operation was considerably more marginal than it had been before the crash.

And then there was Air France to consider.

It was tempting to believe that returning to service – the euphoria that had resulted across BA as we'd regained the aircraft, the shining white symbol that had provided us with our 'halo effect' – would be the same for the French as it had been for us. But things were different across the Channel.

Gonesse, for a start, had been *their* accident.

And the crash had impacted the French psyche in different ways.

Commercially, their operation had never been as successful as ours. Their average load factor across the Atlantic had been around 35 per cent; ours had been around 57 per cent (and on many flights, we were full to capacity).

There were the differences in our operations too – they flew a 'business-plus' service; ours had always been pitched at 'first class-plus'.

And topping it all was the difference underpinning the arrangements between the two airlines and their – our – respective governments. The Lord King deal that had given us full control over Concorde's profit-and-loss had never been emulated on the other side of the Channel. In French government eyes, and to a degree in Air France's, Concorde was, and always would be, a liability, because she would never be able to recoup her original investment.

In the prevailing mood, the publication of the final version of the BEA report was never going to act in the aeroplane's long-term interests either.

The final addenda to the BEA report were published in January 2002.

Given the UK's experience of the investigation process, the 'Probable Causes' came as no great surprise.

The accident, the report said, was due to the 'high speed passage of a tyre over a part lost by an aircraft that had taken off five minutes earlier' – the Continental Airlines DC-10 – 'and the destruction of the tyre'. It went on to describe the 'ripping' (sic) of the fuel tank in 'a complex process' involving a 'transmission of the energy produced by the impact of a piece of tyre at another point on the tank'.

It then went into a discussion on how this transmission of energy had caused an associated 'deformation of the tank skin and the movement of the fuel, with perhaps the contributory effect of other more minor shocks and/or a hydrodynamic pressure surge'.

The fire, it concluded, had been caused either by an electric arc in the wiring in the landing gear bay or through contact with the hot parts of the engine 'with forward propagation of the flame causing a very large fire under the aircraft's wing and severe loss of thrust on Engine 2, then Engine 1'.

Finally, it acknowledged the impossibility of retracting the landing gear and how this probably contributed to the 'retention and stabilisation of the flame throughout the flight'.

No mention in the 'Findings' section of the possibility of the forward fuel tanks having been filled beyond their mandatory 95 per cent level.

No attachment of any great significance to the crucial shift in the centre of gravity.

The estimated weight of the aircraft on departure, it also summed up, had been 'in accordance with operational limits'.

And, lastly, barely a word of criticism for the crew's actions when AF4590 had reached the end of the runway and called the tower for take-off clearance.

At 14:42 and 17 seconds, the controller had told Marty that they were clear to go, adding: 'Wind zero ninety, eight knots.' To which, Marcot, his co-pilot, had replied, calm as you like: 'Four-Five-Nine-Zero, take-off Twenty-six Right.'

Those nine words epitomised everything that had gone wrong.

In fact, in them, I now saw, lay the keys to the entire disaster.

That night, with the Final Report and its Addenda in front of me, I sat in my office at home and, long after Chris had gone to bed, reviewed again the transcript of the cockpit voice recorder.

At 14:40 exactly, from the tower: 'Four-Five-Nine-Zero, line up on Runway Twenty-six Right.'

To which Marcot responds: 'We line up and hold Twenty-six Right, Four-Five-Nine-Zero.'

'Ready in the back?' Captain Marty asks.

'Let's go,' Marcot says.

There then follows two minutes of discussion as they're held on the threshold – Marty asking Jardinaud about how much fuel they'd used, the flight engineer then giving Marty status readouts on brake temperature, noise reduction parameters and the C of G status at 54.2 per cent. Then, when they are finally ready to go, this from the tower: 'Air France Four-Five-Nine-Zero, Runway Twenty-six Right, wind zero ninety, eight knots, cleared for take-off.'

And then, right back at them, that cool acknowledgement of Marcot's: 'Four-Five-Nine-Zero, take-off Twenty-six Right.'

I pushed my chair back from my desk and rubbed my eyes.

As part of their pre-flight preparations, the non-handling pilot – in this case, the co-pilot (if we were adhering to the BA way of doing things) – would have done a careful calculation of their take-off weights and speeds.

To compute this accurately, he would need to know the aircraft's exact weight, the airfield's height above sea level and surface temperature, plus wind speed and direction. This calculation would take place about thirty minutes before engine start-up using the conditions that prevailed then. If they changed close to take-off, then they'd have to be updated accordingly.

All these data were necessary to work out whether it would be safe to take off on the basis of an assumed engine failure at V_1 – 'decision speed'.

V_1 – the point at which you decided whether to stop the aircraft on the runway using brakes and reverse-thrust or get airborne with enough height and speed to clear a

CONCORDE

thirty-five-foot obstacle at the end of the runway and accelerate to V_2 engine-out climb speed with the undercarriage up – was virtually tattooed on the forehead of every pilot from the earliest days of training.

V_1, as every commercial pilot knew from the very first, was not a fixed number necessarily – it was a number in a band. But it was predicated on the computation that the non-handling pilot made before take-off. This translated as: *at what weight can this aeroplane, on this runway, on this day, in these conditions, get airborne and satisfactorily climb away on three engines?*

Charles de Gaulle had a very long runway and, despite the fact it was mid-summer, there had been a sufficient margin for the aircraft to take off.

Concorde, like other aircraft, had two take-off weights: its 'maximum take-off weight' of 185,070 kilos – the weight, specified by the manufacturer, above which it should never take off; and what was called the 'performance-limited take-off weight', which was the number that could vary, depending on the circumstances of the day.

The 185.070-tonne maximum take-off weight figure was specified by the manufacturers and authorised by the regulators. It was there to ensure that certain stress levels were not exceeded to preserve the aircraft's long-term structural life. Whilst you definitely weren't meant to exceed this, if you happened inadvertently to do so every now and again by some small margin (not that you'd ever intend to), so be it; the aircraft wasn't going to break.

The performance-*limited* take-off weight, on the other hand, was an *absolute limit* determined by the on-the-day, at-the-moment circumstances. If you exceeded it, and you had an engine failure, the maths said very simply you wouldn't clear that 35-feet obstacle at the end of the runway. Of course, if you didn't have an engine failure, you'd be fine. But aviation doesn't work like that. You plan for what *could* happen – not what you hope will happen.

The manual, logically therefore, said you should observe whichever of the two limits was the more restrictive on the day.

Most of the time we were limited by the max take-off weight, 185.070 tonnes, because most of the time, we operated out of big airfields with long runways – airports like Charles de Gaulle. Occasionally, however, we encountered hot days, with no wind, or maybe even a tailwind, where the performance-limited take-off weight became the limitation because it dropped below the max take-off weight.

The pencil and paper calculations for a Concorde flying out of CDG under the conditions initially used before engine start-up on 25th July 2000 were well above the max take-off weight – in the order of 188 tonnes. And the signs said that *this*, on that fatal day – in contravention of the rules; in contravention of all *sense* – was the weight that they were really working to.

Why did this matter?

A computer, as we all knew, as everybody knows, is only as good as the information it receives: *garbage in, garbage out*. Concorde's computer was no different.

On that day, the load computer thought that it had all the fuel on board that it was supposed to have – no more, no less. As far as its computational algorithms were concerned, it had burnt its taxy-fuel and hadn't had its tanks filled beyond 95 per cent. And, because it believed the aeroplane was below its never-exceed performance-limited take-off weight, it had calculated that, were it to have an engine failure at V_1, it would be

OK – it would still be able to avoid that 35-feet obstacle at the end of the runway and climb away safely at V_2.

That would have been all well and good – not right, but as may be – were it not for one thing: they'd worked out the performance-limited take-off weight on the basis that there was no wind – they'd be taking off in still air.

'Air France Four-Five-Nine-Zero, Runway Twenty-six Right, wind zero ninety, eight knots, cleared for take-off.'

'Four-Five-Nine-Zero, take-off Twenty-six Right.'

In their take-off calculation, Marty's crew had assumed *zero* wind speed. The tower, though, had just given him the wind speed – taken from anemometers deployed by the runway – at the moment of take-off as 'zero ninety, eight knots' – meaning easterly at 8 knots or, in plain language, 'behind you at 9 mph'. If you'd been standing in it, a noticeably 'gentle breeze'.*

This mattered because, suddenly, this was more *limiting* than the original factors that had been baked into the performance-limited take-off calculation.

Nine mph is appreciably more limiting than zero.

In effect, a tailwind blowing down the runway behind them *extended* their take-off run, *reducing* their margins. Had it been a headwind, blowing *more* air across the aircraft's lifting and control surfaces, it would have increased their margins. But it wasn't. In the calculations you factored into the performance-limited take-off calculation, to err on the side of caution, you always allowed for half of the reported head-wind and one and a half times the tailwind (i.e. if there was a headwind of 8 knots – assume 4. If there was a tailwind of

* As it's described on the Beaufort Wind Scale.

8 knots assume 12. The computer knows this and will do that adjustment for you).

Consequently, the impact of the tailwind was massive. In AF4590's case, at a stroke, it recalculated the performance-limited take-off, which had been about 188 tonnes, down to an infinitely more critical weight of around 180 tonnes, way below the max take-off weight – and the performance-limited weight that they had assumed. Expressed the other way round, not only had they planned on taking off considerably above the maximum take-off weight, but they'd also planned on getting airborne above the never-exceed value of the performance-limited take-off weight, the one thing you never, ever did.

When AF4590 turned on to Runway 26R, its real weight was around 186.5 tonnes – very likely more because of the 'unknown' quantity of bags and newspapers aboard. That was a tonne and a half over the maximum take-off weight *and all of 6 to 8 tonnes over* that never-exceed performance-limited take-off weight, having been told that they had an 8-knot tailwind.

An 8-knot tailwind that they never even discussed . . .

In that moment, in the dark of my office, the BEA report lit only by the light of my computer in front of me, the feeling I got from it almost overwhelmed me.

The picture I'd built to this moment of a crew under pressure to get away – the rushed checks, the 2 tonnes set aside for taxying with only 800 kilos used, all the various missed tech calls – to a degree told its own story.

But reading through the report – focusing on its details – had triggered something else: a part within me that had been engrained from three-and-a-half decades of life lived in a cockpit; longer if I took into account the passions of that small boy who had gazed up at those silver specs from the beach.

This was the part that Marty and his crew had glossed over – the approach they had taken to *flying*.

It was in my nature to be a stickler for the rules, but it was also instilled in me as a pilot – and in my position as the boss of the Concorde fleet at BA. I was the person who not only *had* to stick to them but know them inside-out.

What I saw in the BEA report – and what I read again now in the transcript of the CVR – shook me. I was, in effect, listening to a crew that was about to die in the most horrendous circumstances along with 110 others. I was listening to the last words of contemporaries and colleagues. But in addition to the sickening feeling of knowing what was about to happen, I was simply staggered – *dumbfounded* – by what I saw:

'Air France Four-Five-Nine-Zero, Runway Twenty-six Right, wind zero ninety, eight knots, cleared for take-off.'

'Four-Five-Nine-Zero, take-off Twenty-six Right.'

No reaction from the crew. No discussion. Nothing.

Thirteen seconds later, Marty had banged the throttles forward and, thirty-three seconds after that, they'd hit V_1: *no going back*.

In the way that we calculated take-offs, wind, along with temperature and altitude, was one of the three critical elements to take into account. Temperature because hotter air is thinner. The altitude of the airport because higher airports also have thinner air. And, finally, wind, which, ideally, you want blowing *at* you – air already moving back across the wing helping to get you airborne.

But because you don't always get the conditions you'd like,

you have to do your take-off calculations accordingly – weight being the most critical factor.

When the crew of AF4590 chose not to discuss, let alone acknowledge, the fact that the wind had changed a moment before take-off, what they also ignored was the fact that, at that same moment, their aircraft, in effect, had become 6.5 tonnes heavier. Or, expressed differently, it needed to lose 6.5 tonnes to take off safely – 'safely' meaning continuing to fly in the event of an engine failure at V_1.

All pilots know this – it's drilled into you from your very first day in the classroom.

The crew of AF4590 would have been no different. They'd come up with a set of figures on that day for the safe performance of their aircraft at take-off from Charles de Gaulle.

Their calculations had assumed zero wind, but – Murphy's Law – they got to the end of the runway to learn it had changed: they had a tailwind.

And nothing was said about it. Not a thing.

What had their options been at that point?

On being told they had an 8-knot tailwind, having planned for zero-wind, they could have sat at the end of the runway and waited for it to drop.

Another option would have been to sit there and burn fuel until they'd got their weight down to their performance-limited weight figure. But that was 6.5 tonnes' worth and would likely have taken the best part of an hour.

Valuable, even essential, time lost – and, in their eyes, they needed every drop to get their fully laden aircraft to New York.

Option three would have been to have taxied to the other end of the airfield and taken off the other way – into the wind. Time and fuel lost again. They would have then had to carry out a technical stop somewhere to *re*fuel. But that would have defeated the object of the exercise, too. Even more time lost – the passengers would have missed the boat.

Next, they could have taxied back to the ramp and taken some of the fuel – some of the weight – off. But, if they'd rejected the other options, this one didn't make any sense either – the point was to *hang on to* the fuel they had.

Or, finally, they could have gone back to the ramp and taken some of the baggage off – or some of the passengers. But that would have taken time and would have pissed *everybody* off. Passengers minus their baggage might have made the ship but would have spent the next two weeks with nothing to wear.

In the end, under severe time pressure, they'd chosen the only other option:

AF4590, clear for take-off . . .

I looked up at the wall beyond my desk. On it were photographs depicting some of the many milestones of my career. A gleaming blue, white and silver BOAC VC10 with its associated memories of Geoff Morrell, my instructor-pilot, Arthur Winstanley, the veteran flight engineer, and Chopper Knights, who, at heart, hadn't been nearly as terrifying as his reputation. All those incredible characters who'd slid out of the left-hand seat to give way to pilots like me: the first generation trained on big jets in the jet age.

In the low light, some of the photos morphed into less welcome memories: Speedbird 712 *Whiskey Echo*, the BOAC 707 that had lost an engine – *physically* lost an engine – after take-off from Heathrow; Speedbird 911, the BOAC 707 that had broken up over Mount Fuji; and the Qantas 707 against which we'd duelled in Mauritius – until it removed the top of a tree.

All of them had held pointers – clues we ignored at our
peril. If Marty and his crew had discussed the fact they had
an unaccounted-for tailwind at the moment of take-off, one
of them might have reminded the others that things
happened – bad things – when you flew outside of an aircraft's limits.

But the incident that came to me most acutely then was
the crash of *Papa India*: the BEA Trident that had deep-
stalled into the King George VI Reservoir at Staines when
one of the crew – most likely, the captain, Stanley Key, who
had strapped into the left-hand seat having had an argument
with another colleague just before entering the flight deck –
had commanded the premature retraction of the aircraft's
leading-edge slats, control surfaces designed to keep an aircraft stable at slower speeds. In the inquiry that followed,
investigators had determined that Key's two junior crew
members may have been too awed by his rank and
experience – Key, too, had been a combat pilot during the
Second World War – to have challenged him. As a result, two
memorable measures were introduced, one technical, the
other procedural, that were supposed to eliminate the possibility of this kind of thing happening again.

The legacy technical measure was the introduction of the
cockpit voice recorder, which would henceforth give investigators a clear understanding of flight deck interactions in the
wake of an accident. The second had been the introduction
of 'cockpit resource management' – later called 'crew
resource management'. CRM was the training that all commercial pilots received to allow them to overcome the
perceived problem in *Papa India* – something known as the
'seniority gradient'. Rooted in CRM was the idea that anyone
could challenge anyone else's decision on the flight deck, as

long as it was done in the right way; and the training was designed to ensure that it was.

Embedded in CRM was the foundational idea that each member of the crew monitored what each of the others did – and, before any major actions were carried out, that those actions were – *are still* – discussed and agreed.

The essence of a smooth-running flight deck is that it's a team game: one person calls for something to be done, another physically does it, a third (if there is one) monitors it's been done correctly.

If something comes up that's unusual – some kind of unserviceability, a passenger who's been taken ill, or a change in the weather conditions – you discuss it as a crew, you plan it as a course of action and then you agree it.

None of these things happened at the critical moment – the moment the tower signalled AF4590 had a tailwind.

The BEA accident investigation fudged this, as it had so many other things, by saying the 'zero-nine-zero at eight advisory' was only a snapshot at that particular moment in time – and this was true; it was. Subsequent to the accident, they looked at all the anemometers along the runway, which were showing different readings (which they would have done in relatively light air) and came to the conclusion that if you were to average all those readings out, you'd have ended up with zero wind. *De facto*, it had been OK for them to take off.

Even if this was true, it wasn't the point. None of that knowledge existed at the time because the nearest windsock the crew could have looked at was around a kilometre away. In this vignette, I – all of us on the UK side of the Channel – were afforded a glimpse of a much bigger picture: one that showed, tragically, that this hadn't been some snap thing – it had been part of a generalised pattern of indiscipline.

AF4590 had been an accident waiting to happen – perhaps for years – only not in the way the BEA report had alleged.

In an addendum to the report, the UK Accredited Representative to the investigation of the crash of AF4590 submitted what amounted to an official complaint about 'the manner in which the French judicial authorities affected the technical investigation'.

In all my years of flying, I had never heard of such a thing.

Among the objections it raised was the way the judicial investigation had been conducted, which, it said, had presented 'major impediments to the AAIB's participation in the technical investigation'.

These included the fact that it hadn't been allowed ('except very briefly') to examine the wear-strip that burst the tyre, or a part of Tank Number 5 that had been found on the runway, or key flight deck controls and instruments.

It also highlighted the French judicial authorities' obstruction of the UK AAIB investigators' systematic involvement in the examination of the evidence; in a word, the UK's representative was saying the UK had been obstructed, pretty much at every turn.

The corollary to these and other impediments, it wrote, was the serious charge that the AAIB had been 'significantly hindered (in) the prompt examination of the evidence' and that this had 'introduced significant delays to necessary safety actions'.

It added that these actions contravened obligations under the Chicago Convention and its European counterpart: that 'investigators should be able to complete their tasks unhindered'.

The BEA, for its part, said it regretted the 'difficulties encountered' – and I had no doubt that it did.

The villain of the piece, as I and others this side of the Channel saw it, was the system itself, which precluded what all air accident investigations, in whatever country, were meant to seek – the thing I had come to think of as 'Amy's wish': that nothing like this should ever be allowed to happen again.

37

Around the time we'd got Concorde back into the air, we had a management reorganisation.

Under the *status quo*, each fleet had had a senior manager who reported to a 'Chief Pilot' for a particular group of fleets, of which there were over a half-dozen. Under the new organisation, they wanted a flatter management structure, reducing reporting lines and costs. There would be a general manager for the long-haul fleets, which comprised the 747 and the 777, and a general manager for all the other fleets, made up of the 737, 757, 767, the Airbus A320 and its family – and Concorde.

I was asked if I'd like to take on the short- and medium-haul fleets and I leapt at it. It was a promotion – up two grades from my 'alpha' management level – but, crucially, it contained a proviso that allowed me to continue to fly Concorde.

I immediately appointed separate flight managers – one each for Concorde, the 757/767 fleet, as well as for the 737 and the A320. This made my job all the more interesting. All of a sudden, I was focused on a swathe of issues beyond Concorde on her own. I was responsible for 135 aircraft and 1,350 flight crew with an operating budget of £135 million. And every fortnight or so I still got to fly the aircraft. It was, in many ways, my dream job.

At the beginning of 2002, we were approached by the organisers of the Queen's Golden Jubilee celebrations – a series of commemorative events planned for June of that year to mark Her Majesty's fifty years on the throne – to see if we would be open to contributing a Concorde to a large formation of RAF aircraft that would fly down the Mall and overfly Buckingham Palace as part of the four-day celebration's grand finale. The request had come from Lord Stirling, Chairman of the Queen's Golden Jubilee Trust, to Colin Marshall, BA's then-chairman.

The decision to accept was by no means easy – the BA board had decided back in the 1980s that it would no longer agree to Concorde taking part in air displays. There had been two exceptions to this ruling in the years since. The first was a commemorative event to celebrate fifty years of Heathrow Airport in 1996. The second was an Edinburgh and Glasgow overflight for the state opening of the Scottish Parliament in 1999. Both were in formation with the Red Arrows, the RAF's world-famous display team. The 'Reds' are a ten-ship team, nine display aircraft with a 'hot spare' that's often also used as a photographic platform, which operate BAE Systems Hawks (originally designed as the Hawker Siddeley Hawk), the RAF's advanced jet trainer since 1976.

From a PR point of view, BA was very keen to participate in the Golden Jubilee. From a flying perspective, however, there was a lot to consider. For our part in the display, we were being again asked to fly in formation with the Reds. And whilst formation flying was second nature to them, it wasn't to us; as airline pilots, we spent most of our lives trying to stay away from other aeroplanes. Consequently, there weren't that many pilots within the airline with that kind of flying experience.

As I'd flown with them for both the Heathrow and Edinburgh events, when the board finally acquiesced to the request, it fell to me to coordinate our participation, personally to keep the board regularly and fully up to date as well as to fly the aircraft on the day.

Knowing what the route would be, we had plenty of time in which to practise – and our simulator was perfect for all the early work. The tricky part, however, was the formation flying, which couldn't be practised on the sim; where nothing short of real flying would do. How, then, to present ourselves on the day? On the one hand, flying slowly down the Mall, the arrow-straight road from Admiralty Arch to the Palace, would deliver real value for money – no one would be able to say they'd missed us when they blinked. On the other, Concorde looked her best with her nose and visor up. This, though, came with certain terms and conditions, one of which was a minimum airspeed of 250 knots. This minimum speed for flying with the nose raised came close, fortunately, to the Reds' optimum display speed – around 300 knots. Being single-engine aircraft this would give them the energy they needed to plan for contingencies should any of them suffer an engine failure.

We compromised, thus, on a speed of 280 knots down the Mall at a height of a thousand feet.

Next, came a debate about whether the Reds should lead us or vice versa. This same discussion had been held before the Heathrow and Edinburgh events.

Prior to 1996, whenever we'd flown together, the Reds had led. In 1996 and 1999, however, we swung it the other way – we led *them* after winning an argument that said they had the formation-flying experience, we didn't, and it would be easier for them to formate on us, rather than the other way around – the Reds by now having learnt how to avoid the powerful and unusually turbulent vortices that spiralled off our wingtips. This had led to an arrow-like formation in which four of their aircraft hung off each wing, with one, the leader, just behind, on our centreline.

This argument, thankfully, won the day for the Golden Jubilee event. It was, of course, entirely coincidental that, from a BA perspective, it looked better if we were at the front!

Having agreed the basics, we held regular meetings with the Reds in the run-up to the full dress-rehearsal. This was scheduled to take place five days before the real thing, which was set for the 4th of June. To assist them, the Reds discreetly marked up the Concorde we'd use on the day with little Day-Glo strips of tape on the fuselage, which would serve as their line-up markers for both the practice event and the Jubilee itself. By far the most time, though, was spent on the paperwork and bureaucracy required to allow us to fly at low level over the centre of London. There had then followed a mild tussle with the organisers over a request by them to inscribe 'God Save the Queen' on Concorde's underside. I was initially quite keen on this but, in the end, we convinced them this was impractical as taking the writing off again would take the aircraft out of revenue service for too long. At long last, we were set to go and practise.

The crew would be a flight crew only. I decided, too, that this would be another opportunity to give the left-hand seat to Jock Reid, the CAA test pilot who had been so instrumental in working with us to get Concorde back in the air. As I was qualified to fly from both the left- and right-hand seats, this meant that we would be able to share the flying on the day.

The practice day arrived soon enough. The formation was to be comprised of a number of different RAF elements: a giant C-17 Globemaster transport aircraft, a TriStar tanker-transport and two Tornado GR4 strike aircraft; an E3 Sentry airborne early warning aircraft escorted by two Tornado F3 fighters, a VC10 tanker-transport accompanied by two Jaguar ground-attack aircraft, a Nimrod maritime patrol aircraft and two Canberra reconnaissance platforms; a BAE 146 of the Queen's flight flanked by a pair of HS125 CC3 navigational trainers; a lone Eurofighter Typhoon; and, finally, the Red Arrows and us.

As the C-17 was the RAF's brand-new toy, it was agreed that it would lead the formation and carry out the navigation for all the other elements.

To ease the bringing together of the entire formation, it was decided to split it into two packages: the Reds and us comprised one package; all the other aircraft comprised the other. It was done this way because being a ten-ship, as well as the finale, we would need to come together as one unit. The Reds mounted out of RAF Manston, and we from Heathrow. We met up over the northern North Sea; the other elements over the southern North Sea. Once the two groups had achieved their formations, the next trick would be bringing the two together *en route* to London, the Mall and Buckingham Palace. To maintain the visual impact on the day, on the practice day, the base at RAF Marham on the north Norfolk coast substituted for Buckingham Palace. The brief to each of us in the formation was simple: follow the aircraft in front of you. The aircraft planned to be in front of us was the RAF's shiny new multi-role fighter, the Typhoon, but it was so new that it was still in flight-test, and couldn't be spared for the practice, so a Tornado GR4 acted as its stand-in.

Everything went like clockwork. The two formations began their run-in towards the coast over the North Sea. The place chosen for the point at which we merged was the lighthouse at Southwold, on the Suffolk coast, 10 miles south of Lowestoft. The lighthouse, as well as being a very clear landmark, just happened to sit on an extended centreline from the Mall and Buckingham Palace roughly a hundred miles to the south-west. The C-17 hit the marker bang-on and everyone else fell in behind, at the right height and speed.

Instead of heading for London and the Palace, however, we turned north for Marham, where an RAF air vice mar-shal was waiting to mark our performance. If we did it right, we'd get the authorisation we needed to do it all for real in five days' time. Fortunately, we did and acquired the piece of paper.

The only other thing that remained for me to do was to overfly the route in an RAF Gazelle helicopter from Northolt. This was laid on to familiarise me and some of the other pilots with the view we'd get from a thousand feet over the centre of London. I noticed as we flew over the City – London's finan-cial Square Mile – that the northernmost tower of the Barbican Centre acted beautifully as a final run-in marker, some three-and-a-half miles from the Mall's extended centreline.

Upon landing, with butterflies in my stomach I'd not felt since the early days of my pilot training, I told everyone who was involved that we were ready.

On the day, the weather was fine – a high cloud base with good visibility – which came as a relief.

Jock and I took off from Heathrow and headed out towards the North Sea, where we found and met up with the Reds. To give the southern North Sea formation elements time to hook up with each other, we began to stooge around the northern North Sea in a racetrack pattern, much as we would were we in an air traffic holding pattern waiting for clearance to land at Heathrow, only we had eight shiny red Hawks off our wingtips and one trailing behind.

We were all set to begin our run-in to the lighthouse to merge with the other formation prior to our heading inland to London, when we got a radio message: Her Majesty had been delayed by thirty minutes. We'd need to cool our heels for a while.

This wasn't a big deal for us, because we – that is, the Red Arrows and us – had plenty of fuel. It was, however, a big deal for the southern formation, several of whose aircraft were critical on fuel. Fortunately, the RAF had thought of this, and several air-refuelling tankers were on standby over the North Sea to top up any of the elements that were running low on gas.

While all this was going on, I called Chris, who, like the rest of the nation, was watching the celebrations unfold on TV. 'We're running around thirty minutes late,' I told her. 'Time for you and Amy to go make a cup of tea. How's it looking down there?'

'You cannot believe how big this thing is,' she replied. 'I've never seen so many people in my life.' I could hear the emotion in her voice. Everybody had been feeling emotional thanks to the four-day, televised build-up. An Elizabethan reign that had spanned a half century. For two-thirds of the country, maybe, Elizabeth was the only monarch they'd ever known.

A crackle in my headset presaged the news Her Majesty was ready.

Jock, who'd been flying up to now, handed me the controls. As we swept in towards the lighthouse at Southwold, we spotted the other formation and began to close in on the back of it. All we had to do now was follow the RAF's brand spanking-new Typhoon, the next element in the formation ahead, and . . .

'Where is that bloody thing?' Jock said, interrupting my thought-flow, as our several-mile long formation prepared to bank inland.

I was craning my neck for some sign of Britain's brand-new multi-role combat aircraft, but all I could see were three tiny specks in the distance – the BAE 125 navigational trainer jets and the BAE 146 of the Queen's Flight.

'No idea, Jock,' I replied.

If any elements had a problem, our brief was to close up on the one immediately ahead, which I started to do, inching the throttles forward until the 125s and the 146 began to fill a section of the windshield.

As we approached Southwold's lighthouse, the Typhoon suddenly materialised, a Tornado off each wingtip. It had apparently suffered a problem with its super-sophisticated navigation suite. I couldn't suppress a slightly mischievous smile. 'I think he must have got lost.'

We turned again for Romney Marsh, on the final leg of our run-in towards the capital. The Typhoon's two shepherds peeled away, leaving the RAF's new fighter tucked in safely between us and the 146 and the brace of 125s. But within a few minutes it began to drift off-track.

'Where in God's name is it going *now*?' Jock said.

The Typhoon seemed to have developed a mind of its own.

Before long, it had drifted so far to the north that it had disappeared completely into the haze, leaving us with a dilemma: did we close up on the next element or line up on the Mall by ourselves?

We'd agreed with the Reds that our last chance for a course

correction would be the Barbican Centre, and we were already practically on top of it.

I adjusted course as best I could, and the Hawks followed. Then I looked up from my instruments. At three hundred metres' altitude – a thousand feet – I was stunned by what I could see: so many faces, people waving flags and umbrellas; the Mall dead ahead, the Palace at the end of it.

And . . . *blast* . . . we weren't quite running dead-centre, as we'd planned and practised, but a dozen yards to its north. It was too late now to do anything about it. And anyway, I was too mesmerised by what was happening in front of me.

The Mall . . . packed.

All the parks and bridges . . . packed.

One-and-a-half-million people, their faces angled skyward – a sea of humanity in one place, at one time and for one purpose only: to celebrate the Queen's Golden Jubilee, all eyes on the flypast – and us as the grand finale.

To my amazement, I could clearly pick out the balcony of Buckingham Palace, adorned with a golden crown on a blue banner. And, front and centre, Her Majesty in a flame-orange outfit. Flanked by her entire family. Their faces angled skyward, too, at *us* . . .

A crackle in my headset. The voice of the Reds' team leader: 'Smoke on. *Go!*'

I couldn't see the red, white and blue trail from the Hawks' exhausts, but something – a frisson, a blast of energy, I didn't know what – seemed to emanate from the crowd, permeate the cockpit and lift us.

For a second, I was my seven-year-old self again, staring at specks in the azure, telling my parents *that* was where I was going to spend my life.

But only for a second.

As we were running in towards the Mall, Trevor Norcott, the flight engineer, had armed the afterburners by leaning between Jock and me and flicking up the four white panel switches at the back of the throttle quadrants.

No carefully graduated sequence today – the two inboard engines first, the two outboards three seconds later for when we went supersonic; the muted, sequential double-kick that ensured we didn't spill the passengers' drinks.

Today, we didn't care – we *wanted* those four Olympus 593s, each generating 38,000 pounds of thrust with after-burner, to give us everything they had.

As we zoomed towards the Palace, I primed the RT.

'Reds, Concorde, break in five . . . four . . . three . . . two . . . one . . . *now!*'

As soon as we were clear of the all-important balcony, I banged the throttles open. The power went through 75 per cent, the reheats engaged and over 150,000 pounds of thrust kicked in, punching me against the back of my seat. Lots of flame, lots of noise as I eased back on the stick and that deep rumble I knew so well reverberating in my chest.

The Reds peeled away in formation below us and we pulled up, up, climbing like a rocket as I'd done on my first Concorde flight, up towards 3,000 feet in the blink of an eye – only, this time, I was twenty-five years wiser, and ready for her.

With a lump in my throat, I thought of my mum and dad again; I thought of Chris and Amy; I thought of the extraordinary effort that had brought Concorde into being; of the people who had maintained her, the passengers who had flown on her and the people in the street who always stopped and looked up. And I thought of the 113 who had died at Gonesse too.

This was for them.

Six months after the successful conclusion of the Return to Service programme, today was the day that Concorde reclaimed her status as an icon.

38

It was at a party just before the end of the year that I got my first hint of trouble. The occasion was our Concorde Christmas get-together, a black-tie dinner-dance affair at a golf club near Heathrow; the moment – as Chris and I were preparing to leave. A guy I knew, a middle manager on the fleet, took me to one side and said: 'What are all these rumours I'm hearing about Concorde retiring, Mike?'

Up to this point, the discussions he'd been referring to were secret – only a few people were supposed to know about the conversations that had taken place regarding the possibility of Concorde's retirement. I'd been sworn to total secrecy – I'd not even told Chris. But the secret appeared to be out.

Even though I had been privy to the discussions, initially I'd not been too concerned about the outcome. These kinds of conversations happened a lot – and not just about Concorde. We were always looking at contingencies. Most of the time they resulted in a study or two, somebody then did a bit of maths to see how the retirement of Fleet X or Base Y impacted the bottom line, then, nine times out of ten, nothing happened. In my mind, the Concorde retirement discussions fell into this bracket – as part of a range of options BA was considering as it looked out across the first couple of decades of the new millennium.

The fact that we were discussing the aeroplane's retirement at all was because we'd been picking up vibes that Air

France was looking at retiring *its* Concordes and we needed to know, were this to happen – which, all in all, to begin with we doubted – how it would impact our Concorde operation.

Whatever the French did, one of the reasons I felt it unlikely *we* would be retiring was because we'd spent so much time, effort, energy, passion, love and money getting the aeroplane back into the air after the accident; and since then, she'd had almost fourteen months of successful operations. But here, now, confronted with the question from someone who had no business to know, I felt a slight, sudden rill of anxiety in my veins – and perhaps, too, as I replied with all the confidence I could muster, the presence of the Marmite anti-brigade.

'As far as I am aware,' I told him, 'the Concorde operation is entirely successful.'

This answer was as true as it was evasive. Concorde was wholly successful – at BA, it was making money, good money. But it wasn't BA that I was worrying about as Chris and I headed for home at the end of the evening.

As the days drifted into the first weeks of 2003, it became more obvious that Air France was considering retiring the aircraft. All the old grumbles seemed to be resurfacing across the Channel. The fact they had a big paper loss on their books. That their 'super-club' service wasn't attracting the same high net-worth traffic as our 'first class-plus' service. Its looming privatisation and the change in the law that would soon see directors accountable for corporate screw-ups.

And last, but not least, the ghost of the accident they had never quite shaken off.

Any chance we might have had of deflecting this fresh set

of whispered misgivings from across the Channel was itself deflected by an incident involving an Air France flight from Paris to New York on 19th February. Even though it was minor – the in-flight shutdown of the Number 3 engine due to what was subsequently shown to be a ruptured fuel line – it had necessitated a precautionary landing in Halifax, Nova Scotia with not very much fuel. And we'd heard it had put the wind up some of Air France's senior directors; their view being that they'd just dodged a bullet, and next time they mightn't be so lucky.

French pressure on BA to go for a bilateral Concorde retirement stemmed, we thought, from their concerns we'd invoke the clause in the treaty that said any unilateral withdrawal – never mind it was now forty years since the contract was signed – would involve compensation to the other from the withdrawing party.

As we were only too aware, things were very different for us Brits. After the 1984 Lord King restructuring deal, in which we'd assumed responsibility for our share of Concorde's costs as well its profits, the costs – related to engineering as well as infrastructure and support – had risen to £60 million per annum. Whilst we were still able to turn a profit on this, the margins had slimmed considerably, and word had recently reached us that Airbus – the aircraft consortium made up of French, German, UK and Spanish aerospace companies – was about to raise the support costs again by around £40 million. Because we had seven aircraft and the French five, this would put us in line to cough up an extra £24 million a year that the operation really couldn't sustain – unless the French honoured their part of the treaty and contributed to the hike.

But that seemed unlikely, too. The whispers from Paris

said that the French would make us sing for their share of the costs – that they'd see us in The Hague, the home of the International Court, before they coughed up a sou. Without Air France sharing the support burden, we'd be responsible for £100 million in engineering and support costs on Day One of the new financial year – before a single Concorde was towed out the hangar.

Even so, I was batting for my aircraft. My pitch to anyone in a position of seniority who'd listen was that the business was cyclical and the down-cycle we were still in post-9/11 would come back up soon enough. I even called on a few board members that I'd known from my Jubilee planning days to make that pitch.

But in making it, I knew there were only two outcomes: either it would work, or it wouldn't.

And what if it failed?

If it failed, I had two more options: I could throw my toys out the pram or I could accept my part in what would be our collective responsibility to make the best of a bad job. It didn't take a whole lot of introspection to appreciate that throwing my toys around wouldn't change a thing; and that maybe it was better to be inside the tent than out of it. Being inside would at least allow me to discharge my part in what I saw as the airline's responsibility to see this extraordinary aircraft through to the end of her life; and to play my part in the way that happened.

That was when the thought entered my head: the retirement of the aircraft – because retirement was now inevitable – should be a celebration, not a wake.

By osmosis, it seemed, even the Marmite anti-brigade started to nod in agreement. No, we should not shuffle our seven

Concordes into a corner somewhere and quietly dispose of them. Here was an opportunity to turn a story we knew was going to be received with sadness and disappointment by very many people into a celebration – a celebration of an aircraft that had been a huge success for BA. Back in 1976, when the aeroplane had entered service, even its most celebrated test pilot, Brian Trubshaw – Trubbie – had shared with me his reservations about Concorde's long-term future. Brian and I had been having a quiet pint in a pub near Bristol soon after the realisation had sunk in amongst Concorde's manufacturers that they weren't going to sell 250-plus aircraft around the world – that it was only ever going to two airlines: us and Air France. Then, in 1986, when BA laid on a ten-year celebration of the aircraft's service with us, I found myself talking to Brian again – this time, he was a passenger on a Concorde PR junket I'd been flying out of New York to Bermuda. 'I'm amazed we're doing this, Mike,' he'd said, as he came onto the flight deck to chat with us, 'because if you'd asked me ten years ago what I thought the life of the aeroplane would be, I'd probably have said ten years.'

I reminded him about the drink we'd had in the pub near Bristol and said: 'Well, funnily enough, Brian, I did – and that's exactly what you said!'

That Concorde had now been operating for twenty-seven years was, thus, something not even the experts way back when had thought possible, so there really was something to celebrate here. She was the first supersonic aircraft to carry commercial passengers; the first large, non-military aeroplane to go Mach 2; the first to be able to fly at sustained supersonic speeds in dry, military (non-afterburning) power; the first plane to be equipped with revolutionary fly-by-wire flight control technology . . . the technological achievements alone went on and on and were legendary.

I once had a conversation with Neil Armstrong, the first man to walk on the Moon, about the relative achievements of the Apollo programme versus Concorde. Neil had been guest-of-honour at an awards dinner given by the Guild of Air Pilots and Air Navigators,* of which at one point I had been the Master. The award was a special one to honour Neil's outstanding contribution to aviation and space technology – not just about his Apollo achievements. BA had rolled out the red carpet, flying him over in Concorde. And because I was a Concorde captain, I had been lucky enough to sit at the same small table as him, away from the main throng of 700 other Mansion House guests – Neil, as was well known, being notoriously shy. The chance to talk to him over dinner turned out to be one of the highlights of my career. What he said, too, about Concorde was remarkable.

'From a technical perspective, you know,' he said, 'Concorde was as big a challenge as putting a man on the Moon. A different kind of challenge, certainly, but in terms of innovation in aerospace technology, it was an *equivalent* challenge.'

When one or two people at the table expressed their surprise at this, he said: 'We first went to the Moon in 1969 and stopped going in 1972, but you're still flying four times a day across the Atlantic all these years later. So, arguably, which was the more successful?'

When I was lobbying for Concorde to continue, I let senior executives at BA in on this conversation. I didn't have to remind them that this technical achievement had come at

* Now the Honourable Company of Air Pilots.

a cost. It had been for the nation – UK plc – a hugely expensive endeavour, initially. The country had spent about a billion in Year 2000-equivalent pounds as its share of the aircraft's development cost. Just as Apollo had attracted all kinds of criticism for the level of cash spent putting twelve men on the Moon, Concorde had had more than its share of detractors too – and some of them were likely to resurface, I said, in the celebrations around the aircraft's retirement I wanted BA now to underwrite.

On the plus side, though, whilst much of that money had never been recovered directly, BA had made a decent profit out of Concorde, money that had come back to the UK Treasury in the form of corporation tax. It had generated employment, too. At one stage 50,000 people in the UK had been working on Concorde. It had been a focus for the UK's aerospace industry – a major spur for aviation innovation – at a time when the sector had been flagging. And there had been political mileage gleaned, as well. It was generally accepted by now that Anglo-French cooperation on Concorde had paved the way for the UK's entry into the European Economic Community, forerunner of the European Union.

That was the rational stuff, I told the execs. What was impossible to quantify – but undeniably real – was the emotion people felt for Concorde.

Whilst the Marmite factor within BA was dimly mirrored in the population at large, the numbers in the anti-brigade 'out there' were fewer.

Concorde evoked strong feelings in people. She was an icon – that's what icons do. You could love her or hate her, but nobody I knew had zero opinion on her. It was this aspect we needed to celebrate, I said – and whatever we did, it had to provide for those three elements – the technical, the emotional and, above all, the love that people felt for her. And, as we, BA, had been reluctant initial stewards of the aircraft – to begin with, I reminded the brass, the airline hadn't even wanted her – I felt we owed it to her now to do the right thing. Without Concorde, we would have been another airline. Thanks to her, we'd been an airline apart.

There was going to be pushback, for sure, I said. We'd be roundly criticised for retiring the aircraft, so, whatever we did, we had to ensure the celebration had enough *oomph*, to overwhelm the negativity.

But we could do that, couldn't we, I said, with a hint of mischievousness, to the execs with the power of yea or nay.

We were the biggest, we were the best. We were the world's favourite airline.

We were BA.

39

It didn't take long for the idea of the retirement-as-celebration to gain traction.

The next question was: how to deliver it? It was clear, initially, that we needed to signal to Air France that we would like to coordinate the retirement with them. The original thought was for us both to retire at the same time. For operational convenience, though, *we* wanted to retire *our* Concordes at the end of the summer schedule, which meant October. Having initially agreed to this, Air France then decided not to delay – to retire in May, now just a few weeks away. This was indicative of our two, at-odds attitudes – and the fact that, in the shadow of the accident, there was never any prospect in France that a dual retirement perceived remotely as celebratory would be sanctioned.

Having decided we would each go our own way, the next question was how to announce it.

In BA at that time, we had a network called 'Forum 150' through which the top 150 managers in the company got briefed by the CEO on upcoming developments. It was decided that an extraordinary meeting of Forum 150 on 9th April 2003 would be the right place to announce the retirement to the upper echelons of the company and to follow immediately afterwards with a public release – until I pointed out that 9th April, being the anniversary of the first flight of the British Concorde exactly 34 years earlier, wouldn't be our best move.

So, we switched to the following day.

It was then handed to me to break the happy news.

When the 10th arrived, I donned my uniform to signal the gravitas of the moment. I explained the rationale for the decision with a slide show that backed up the arguments we'd rehearsed in private over the past several months.

As expected, we hit the Marmite factor head-on – roughly two-thirds of the senior management expressed their great sadness at the decision, with the remainder expressing their relief. No one was ambivalent – something we'd anticipated, of course, and which presaged the public reaction that greeted the news when the press release went out an hour or so later.

Still wearing my uniform, I found myself put up as the company spokesman on the issue – I'd worked with the press office for many years, and we'd developed a good working relationship. They knew they didn't need to script things for me. As with any announcement, though, there were one or two major points that we wanted to get across.

For our customers, we wanted to stress that this was a commercial decision, that it had been spurred by factors beyond our control, and that, of course, we'd continue to look after them via the service we provided on our 747s through Club Class and First, both of which had just been upgraded.

Having been so vocal to my peers that Concorde shouldn't retire, it felt OK now, ironically, being the public face of the retirement decision – in part, because I could show the company I was capable of accepting collective responsibility, and also because, now that there was no going back, I wanted to be a part of the process that would see the aeroplane retired with all the dignity she deserved.

A celebration, not a wake.

Even so, facing up to the TV lights at the press conference that day was hard. And when, in the Q&A, that most basic of questions was put to me by a reporter from one of the TV news channels – *why* is she being retired – for a fraction of a second, I hesitated. In that moment, I knew what we needed was a simple expression – a sentence, a mantra, if you will – that captured it.

'It's the right decision at the right time for the right reasons,' I said in a moment when the words just came. I had no idea where they'd come *from*, but they seemed to sum it up – never mind that, for a long time, it had internally seemed to me to be quite the wrong decision at the wrong time for the wrong reasons.

From that moment on, this line became universally adopted as the party line.

Over the coming days and weeks, we fielded a variety of questions on the decision. 'Why did you spend all that money getting Concorde back into the air, only to retire it less than two years later?' was one that came up a lot.

And: 'If the numbers don't add up, why don't you retire it *right now*?'

Both were answered with what was, for BA at least, the truth: retiring Concorde had not been due to the accident; it had been for sound commercial reasons.

The question that required a little more thought was: 'Why don't you let that nice Mr Branson have it? After all, we, the UK government paid for the aircraft, so why not just hand it to him, because he says he can operate it, even if you can't.'

Fortunately, we'd anticipated this.

Richard Branson, the boss of Virgin Atlantic, was extraordinarily good at jumping in at opportune moments and this, clearly, was one of them. If we'd given our fleet of aircraft to him and he'd made a success of it, he'd have said: 'Well, there you are, BA couldn't run a tea shop'. If he'd made a hash of it, which we were fairly confident he would have, he'd say: 'BA didn't look after its Concordes properly and here's the proof!' If we said we were refusing to give our aeroplanes to him, he'd have said that BA were being beastly to him again. The party response to this question, therefore, was: 'Because the aeroplane was originally funded by the UK government, the decision to transfer ownership isn't ours to make.' The press knew it was a bit of a cop-out, but it contained just enough truth to carry the day and, after a while, the question went away. Not before I'd bumped into a senior Virgin pilot at a Guild of Air Pilots and Air Navigators do, however, who said: 'You're not really going to give Concorde to us, are you, Mike?' To which I'd responded, not a chance in hell. 'Thank God,' he'd replied, placing his hand on his heart, 'because it would probably send us bust!' Which it almost certainly would have, because if *we* were struggling to pick up the costs, they would have killed Virgin.

The other question we got – a lot – was: couldn't one aeroplane be retained for high days, holidays and air shows? It was a fair question because this had been done with other retired aircraft types, most notably the Lancaster, Spitfires and Hurricanes of the RAF's Battle of Britain Memorial Flight; and, in a private capacity, more recently, the Avro Vulcan V-bomber.

The answer was, sadly, straightforward. Concorde was a lot more complex than the Vulcan, even. The Vulcan that was flying, post its retirement with the RAF, was on a thing called a 'Permit to Fly'. It had lower certification standards because it hadn't ever been required to carry passengers.

Because Concorde was a passenger-carrying aircraft, it had needed a 'Certificate of Airworthiness' and a 'Type Certificate', which were much higher standards of certification than the Vulcan's permit. To fly at air shows and such like, Concorde would need to retain these certificates, all of which required money – lots of money. And who would have paid? As we knew all too well, the £60 million per annum infrastructure cost of maintaining the aircraft was about to soar by another £40 million. Infrastructure cost didn't shrink with a reduction in fleet size – it was the same whether you operated seven aircraft or one. Even if you were *really clever* and managed to get the cost down to £50 million – a reduction of 50 per cent – it would still amount to a million a week. *For one aircraft.* And even if *that* were possible, who would maintain it? Because it had a Certificate of Airworthiness, there had to be an associated Type Certificate Holder for the engineering, which would normally be the manufacturer, which was now Airbus. And Airbus had made it very clear they weren't interested – this being one of the reasons for retirement. Airbus's new wonder-plane, the A380, was approaching first flight, and its short-haul A320 family was still proving phenomenally successful – the company's focus, rightly, was forwards, not backwards. Technically, another engineering organisation could have taken on the Type Certificate. The Air Operator's Certificate wasn't the problem because we held that. It really hinged on the Type Certificate and with Airbus determined to surrender it, there were no other realistic candidates to take on the role. And even if there were, Airbus was under no compulsion to transfer it – in fact, because so much of Concorde was still so technically advanced, it would have been in the plane-maker's interests *not* to

337

transfer it, along with all the drawings and technical specs that would have come with it. Quite the reverse.

For one thing, the clever workings associated within the air-intake computers that controlled the airflow into the engine right across the flight regime were not only proprietary, but secret – this because the technology was in use on a number of military aircraft that Airbus produced or maintained. Plus, this aside, there was no upside in it for Airbus, because nobody would ever say at an air show, 'Look, there goes an Airbus Concorde!' the only time Airbus would ever have been associated with it would have been if something went wrong.

After the initial flurry of questions over our motives for retiring the aircraft, the temperature of the public relations discussion changed as we drifted from spring into summer – much more to where we wanted the discussion to be, which was about Concorde and her legacy. You'd see it in newspapers and hear it in pubs. People were beginning to talk about the 'wonderful job' UK plc had done in taking on the challenge in the first place; the fact we'd done something the Americans hadn't, and we'd done it well.

I even began to overhear conversations that began with: 'I never got to fly on Concorde myself, sadly, but I know someone whose Aunt Mabel once flew on it and she'd had the time of her life . . .' Bit by bit, the clamour for a celebration of the aircraft's life after twenty-seven years of successful operation in the UK eclipsed the doom-merchants. Yes, BA ensured that the aircraft got all the recognition she deserved, but we also ensured that this would be a celebration of all the *people* who'd ever been associated with Concorde. The people who'd designed her. The people who'd built her. The people who'd brought her into service, operated and maintained her – flight crew, cabin crew, engineers, baggage handlers, ticket agents . . . all of them.

Last, but not least, there was the public – the public that had travelled with us as well as those that hadn't. And what was gratifying – per the conversations I'd overheard in the pub and elsewhere – was that so many of them seemed to know at least something of her story already, whether it was one of those headline performance figures – of the 'faster than a bullet' kind or 'flying on the edge of space' – or the fact that she had been paid for in large part by the British taxpayer.

You began to notice it while walking through London, particularly, when, up to six to eight times a day, the aircraft took off or landed, Olympuses crackling as she roared overhead.

Concorde was not a rare sight and yet people still stopped and looked up after all these years. They always had, but they looked up more now because they knew it would all soon be coming to an end. Edward de Bono, the author, inventor and philosopher, once said to me that this was because she appealed to both sides of the brain – the left and the right, the analytical and the artistic – because she was technologically superb and artistically graceful. That might have been so, but for me the reason was far simpler: at some level, most Brits had individually come to feel a sense of ownership. *That's my Concorde up there – something of which I can be rightly proud.*

40

In the run-up to the retirement, I thought there might be a bit of a rush to buy Concorde seats, coming on the back of the 'round-the-bay' charters we'd flown before the accident, which had been enormously popular and also a money-spinner. I suggested putting the ticket price up. My colleagues in marketing, however, weren't so sure. The prevailing view on their side of the house was that people would see Concorde retiring because she was 'old' and with that came notions of 'unsafe'. The memories of those terrible pictures on the day of the accident had been seared into the public's consciousness – all of which added up to what the market-eers said would be a challenge when it came to selling tickets over the next six months.

For this reason, when they did go on sale after the announcement, a lot of seats were marketed at a heavily discounted rate – you could buy a ticket to New York from London, for instance, for £1,200 as opposed to the normal £4,000.

These seats lasted three days before it became clear that something was going on – that some kind of real, but strange, phenomenon was at work.

Before long, if you could actually get one, a round-trip seat was selling for around £10,000.

But you couldn't get one – that was the point. Before the accident, lifetime average load factors had been around 57 per cent. For all that time, the passenger profile, however, had never changed. It had remained 80 per cent business, with the remaining 20 per cent also unchanged: 10 per cent rich and famous; 5 per cent sports personalities and celebrities; and 5 per cent trip-of-a-lifetime.

As soon as the retirement decision was announced, those stats were stood on their head: 80 per cent became trip-of-a-lifetime; just 10 per cent were business; 5 per cent rich and famous; and 5 per cent sports personalities and celebrities.

Why did it flip? It wasn't that the business demographic didn't want to fly on Concorde – they just couldn't get a seat. Knowing that the aeroplane was retiring, those seeking their trip-of-a-lifetime were booking seats up weeks, if not months, in advance, which wasn't how the business community worked at all. CEOs, COOs and CFOs were used to asking their PAs to book them a seat to or from New York last minute: 'Get me a seat on this evening's Concorde!' But now those seats had all gone. This came as a huge shock to our loyal business customers, until we got a grip on the situation and with a bit of neat sales patter were able to slide them across to our new revamped 747 service.

From this moment on, the atmosphere on board Concorde changed.

There had always been a club atmosphere on Concorde in the best possible sense – those business passengers often knew each other on sight. They were frequently on first-name terms, too, even when they worked for rival firms. Secret deals were done on board. It had all been quiet, low key. A little like the 'right stuff' that existed amongst test pilots, it had been a badge of honour, almost, amongst this fraternity, not to admit to being impressed by it; that this was 'all in a day's work'. But with 80 per cent of our passengers now being trip-of-a-lifetime, all that had changed, Now, it was party, party all the way.

And they wanted it *all* – the Concorde Lounge; the champagne; the pre-flight canapés; the gee-whizz of going supersonic; the thrill of a Mach 2 supercruise; the curvature of the Earth from 58,000 feet and being served the best meals and the finest wines by the best cabin crews in the world. All of the stuff that had been available to our business passengers that they'd largely taken for granted – including the stellar service that met you at your destination (including baggage that was always there before you got to the baggage hall and a process that whipped you through immigration) – was now available to a whole new customer-set and they unashamedly, absolutely, quite rightly, grabbed it all.

As customer service people, we, in turn, responded. Not only did we recognise the new atmosphere, but we encouraged it. For us, the flight and cabin crews, it began to feel an entirely appropriate way to say thank-you to all the people who had flown with us on this remarkable aeroplane over the years. Capitalising on the new mood, we organised 'UK and Overseas Concorde Specials' as mini-tributes to the places that had been so integral to the aircraft and to give even more people an opportunity to fly her. These included Edinburgh, Belfast, Cardiff, Birmingham and Manchester in the UK and Washington, Boston and Toronto in North America. Barbados, of course, figured too.

This all morphed into thoughts on what to do for the final flight.

What, we discussed among ourselves, could we do that would be special? We weren't just going to park the aeroplane, that was for sure.

In due course, the idea of a three-aircraft simultaneous landing at Heathrow – each plane touching down right after the other – began to take shape and emerge.

It was something that hadn't been done before and could not, by definition, ever be done again.

It was also something that we found instant support for when we started to discuss it with other 'stakeholders' – Heathrow and UK air traffic control, for example. But where would – *should* – the three flights come from?

New York was a must. The Big Apple had been central to the whole rationale for Concorde and the relationship between aircraft and city had been special.

Edinburgh, too, felt right. The Scottish capital had been integral to the aircraft's development and manufacture. Edinburgh was also emblematic of the unity – the Britishness – that had underpinned our side of the programme.

The final flight, we determined, ought also to be a salute to the people who had helped to make Concorde a plane of people's hearts and minds; and there was no better way of doing this, we thought, than making the third flight one that represented our trip-of-a-lifetime customer base.

It ought, thus, so to speak, be a 'round-the-bay' flight.

Of the three, the New York flight would be the last to land – Concorde's last-ever scheduled flight.

The next question was who should be on these flights? The idea that took root was that it ought to be a mix – a mix of all the people who'd ever flown with us: business, celebrities and 'ordinary folk', plus people who'd made the aircraft possible – designers, engineers, airline personnel – with, finally, a hint of media. There would also be a raffle – an internal raffle for seats for BA personnel and an external raffle for competition winners amongst the public.

This seemed to take care of it, with one notable exception. Before the retirement decision was announced on 10th April, there were only two people who had actually booked and paid for seats back from New York on the day that we'd selected as the last-ever scheduled flight: 24th October. These people were a retired British couple – I'll refer to them as Mr and Mrs Smith – and, when we checked our records, as luck further had it, they fell into the trip-of-a-lifetime bracket: this was something they had always wanted to do.

It fell to me to give them a call.

'Mr Smith?' I enquired as a man picked up and answered.

'Speaking.'

'This is Mike Bannister from British Airways, Mr Smith. I'm the Chief Concorde Pilot.'

'Oh,' he said. 'Did you say—?' He paused. 'Is everything all right with our booking?'

'Well,' I said, 'it is regarding the trip you booked and paid for on the 24th October from New York that I'm calling. Have you seen today's news?'

'No,' he replied, warily.

'Well, I've got some good news and some bad news. Perhaps I'd better start with the bad . . .'

I proceeded to explain that an hour or so ago, we'd announced Concorde's retirement and that the flight he'd booked him and his wife on to – the only customers to have done so – was Concorde's last scheduled flight.

'Oh, that's terrible, and so sad,' he said, genuinely upset. 'But why?'

I explained the rationale for the retirement decision, which he listened to patiently.

'What's the good news?' he asked, sounding despondent.

I said: 'We're going to give you your money back.'

And, of course, we did. It amounted to around £10,000. But we also gave him his seats on the last flight for free.

41

The final flight came around all too soon.

Our aircraft was Concorde G-BOAG – *Alpha Golf* – one of the two final Concordes to have been built in Britain, and the Concorde with the lowest number of airframe hours. In the early 1980s, *Alpha Golf* had been taken out of service and cannibalised as a spares source, then returned to airworthy status in 1985. In her life, she had gone supersonic more than 5,600 times.

I was part of the crew that had been designated to fly her. To ensure we were fresh and rested for the whirlwind of publicity that preceded it, we all flew out by 747 to JFK on the Monday, with the Concorde flight scheduled for the Friday.

Sitting in the lounge, pre-departure, I spotted Sir David Frost, the well-known journalist and broadcaster.

David was a regular Concorde customer whom I'd come to know a little from his various trips up to the flight deck. Unlike many of the celebs who travelled with us, he never insisted on sitting in the 'celeb seat' – Seat 1A – settling instead for further back; the very back – where he was more likely to sit adjacent to an empty seat and could spread out his multitude of newspapers in preparation for his TV shows. The lore of Seat 1A had come about via the mistaken belief it was earmarked for the Queen. Her Majesty, in fact, sat in her own specially configured seat when she flew with us – two sets of seats facing across a table, all of which had to be installed by Engineering pre-flight.

Normally, I'd have not bothered a customer, but, on this occasion, our eyes met and David waved me over. As we talked, it turned out that part of our respective 'missions' overlapped – he had been booked to do some TV and media interviews over there on his twenty-some years of Concorde travel.

'But the main reason I'm going over is so that I can fly back with you lot,' he said.

'That's great,' I told him, not revealing I was privy to the list of passengers flying with us. David's name had been one of those on the media list who'd risen very quickly to the top. We agreed we'd try and catch up on this flight, but it didn't turn out that way. The next time I saw him, we were in the same New York studio for NBC's morning *Today* show. I was speaking about my role as pilot; he as a Concorde customer of long-standing. The anchorman was less interested in the gee-whizz stats I thought I'd come on the show to provide, more in extracting from me juicy tales of some of the famous types we'd had on board. I tried to deflect him with the names of those who were well known for travelling on Concorde – Taylor, Nureyev, Jagger, Frost – and smiled, as I nodded to David who was at the other end of the sofa.

But the anchor wasn't having any of it – this guy wanted me to spill the beans.

'One of the things that British Airways offers is discretion,' I said, mock-pompously.

Afterwards, he joked that he thought he'd got me, but, like David and his penchant for a quiet, unassuming seat, I had a rule: what happened on Concorde, stayed on Concorde. As far as I was concerned, that rule still held.

On the morning of the flight, we got out to the airport early – around 05:15 – in the pitch-dark. JFK at this time was unfamiliar to us; our normal departure time was 09:30, but today we were off the chocks at 07:00.

After the crazy whirl of publicity over the past several days, it was a relief to get back into 'piloting'. I'd agreed with my co-pilot Jonathan Napier that he would do the take-off, take us supersonic and into the supercruise, and that I'd take over mid-Atlantic and carry out the landing at Heathrow.

There were five of us on the flight deck, unusually – two flight engineers and three pilots; plenty of redundancy in case one of us got sick.

The first thing that greeted us was a gathering of BA staff, many of them in tears.

For these employees, Concorde had been a very large part of their professional lives; for a number of them, it had been all they had known during their time with the airline. This was a small, close-knit group, many of whom had established strong friendships with one another, and with us Concorde crews. They presented all of us on the flight deck that day with gifts they'd clubbed together to buy: cut-glass apples, each engraved with our names. To say I was touched was an understatement. It still sits on my desk.

As we chatted, BA's head of publicity in New York, John Lampl, tapped me on the shoulder and asked if I might say a few words at what he said was 'an impromptu gathering' in a nearby room. The impromptu gathering turned out to be a heaving mass of the great and the good – including Mayor Giuliani and a host of our best customers from the Eastern Seaboard – people who'd come to JFK at this god-awful hour to say goodbye to the aircraft they'd loved.

'Do you mind circulating and, as it were, pressing the flesh a little?' John said.

'OK, but not for too long. I've got an aeroplane to fly.'

John looked crestfallen. 'We were hoping you'd say a few words. You know, make a speech.'

'A speech? Here?' Nobody had said anything about a speech.

'At the gate, in fact.' He glanced at his watch. 'Literally, just before you fly.' It was already almost 06:30 – just half an hour to go.

At that point, I gave in. I found out only when I got to the gate that it was to about 150 people and being beamed to the world live.

'There are three great lady loves in my life,' I said, as I got up onto the dais, beseeching as I did so the gods who helped people who winged it at moments like this, 'my wife, Chris, our daughter, Amy, and Concorde. There's only one of them I can control – you'll be pleased to hear that it's Concorde. And in a few moments' time, I'll leave you and go and start taking the controls of Concorde for this, the final flight from New York to London . . .'*

I reiterated what I'd said elsewhere: that this was a celebration. A celebration of the aircraft and her capabilities. A celebration of her twenty-seven years in BA. And a celebration of all the people who'd ever had anything to do with her.

As for the people who had regularly travelled on her, I said, they had become part of the Concorde family too.

'If you're part of that family already, having flown with us before,' I said, looking at many familiar faces in the crowd, 'welcome back.'

Mr and Mrs Smith then popped into my head, the only people to have booked seats on this flight – they were here

* The event was recorded for posterity here: https://www.youtube.com/watch?v=WbiRz7CbCoE

too. 'If you're new to that family, welcome. We're delighted to have you join one of the most exclusive families in the world. Because in a few moments' time, we're going to take you on Concorde, the world's only supersonic airliner. We're going to take you to the edge of space, where the sky gets darker, where you can see the curvature of the Earth.

'We're going to travel across the Atlantic at twice the speed of sound – faster than a rifle bullet, 23 miles every minute. We're going to travel so fast, we're moving faster than the Earth rotates. And the world will be watching us.'

Thirty minutes later, Jonathan eased *Alpha Golf* off the blocks and headed for Runway 31 Left. As we taxied out, the New York Fire Brigade had decided it was going to lay on its own celebration – a water arch of red, white and blue from three of its tenders.

We sat on the end of the runway for two minutes, engines spooling.

Then, because our departure needed to be timed to the second, ATC started to count us down.

Jonathan released the brakes. 'Three, two, one . . . NOW!'

The reheats banged in and the familiar kick in the back; the rush of adrenaline – *still*, after all these years. We shot down the runway and leapt into the air.

At just 50 feet, Jonathan banked us out over Jamaica Bay in what was one of the most exciting, sensational departure procedures for any airliner anywhere in the world.

I looked at my watch. We were slightly ahead of schedule, which would be useful when we came to coordinating with the other three flights.

Out over the Atlantic, Jonathan got her supersonic and into Mach 2.

Uniform on, cap under my arm, I then went back to talk

to the passengers. In addition to those whom I'd seen in the crowd at the departure gate, other familiar faces pulled into focus: Sting and his wife; Jeremy Clarkson and his nemesis Piers Morgan, Sir David Frost, Sir Colin Marshall, British Airways' chairman . . . the atmosphere was carnival-like and, in keeping with our mantra, we were going to celebrate this for all it was worth.

Back in the left-hand seat, past the flight's halfway mark, we began to pick up messages from the 'Shanwick' and London air traffic control centres – messages of congratulation and goodwill as well as messages of sorrow.

As we closed in on the capital, ATC routed us into the landing pattern that we'd verbally rehearsed: G-BOAE from Edinburgh, G-BOAF from its trip 'round the bay' (of Biscay) and us, landing in sequence after we'd first passed low over London.

At this stage I was flying her on autopilot. Jonathan asked whether he could continue to fly her manually right up to final approach. Of course, I replied, and popped her out of autopilot.

The weather was crisp and clear. A flood of memories of the Jubilee hit me. The crowds. The Reds. Buckingham Palace. The bunting. *Had it really only been the year before?* I tamped down my feelings in order to focus.

As we swung towards final approach, I took back control. We lowered the nose and visor as we passed 270 knots, and selected undercarriage down.

It was tea-time in the English suburbs below, the sun lowering in the west as we lined up on 27R, the northernmost of Heathrow's two runways. And still the messages from air traffic kept coming: 'I was the controller the day *Alpha Golf* flew for the first time . . . we're going to miss you,

Concorde . . . Godspeed, *Alpha Golf* . . . Concorde, it's been an honour and a privilege . . .'

Ahead, I could see the two other aircraft, their wings and fuselages bright against the darkening sky. Each was receiving their own messages of goodwill too.

Using the auto-throttles, I slowed *Alpha Golf* to her final approach speed.

Our attitude came up as we slowed. More power as we reached 190 knots well on the back of the drag curve. I figured today nobody would mind if we were noticeably loud.

Still half a dozen miles out, somewhere over Hammersmith, I peered out of the window to my left and saw something I didn't expect: a sea of faces.

As we flew along the A4, the dual carriageway that pointed unerringly at Heathrow's runways, I could see cars pulled onto the side of the road.

Everywhere people were looking up. One guy had pulled his white van across the carriageway at right-angles to the traffic and was standing on the roof, waving a huge Union Jack. The police had expected there would be a bit of disruption around the airport as we three had swept in, but nothing like this.

'Concorde *Alpha Golf*, for the final time, wind 230 degrees, ten knots. You're cleared to land . . .' A medium crosswind from the left. *Concentrate more. This is the last time ever that a Concorde will land at Heathrow.*

As we came in over the threshold, I uttered a silent prayer – *please don't let me bounce this one.* Close the throttles, kick off the drift and then my thanks as the main wheels touched down smoothly. A bit of back pressure on the stick, to 'fly' the nosewheel down, grease the tarmac, reverse-thrust.

I braked for the last time and swung the aircraft onto the taxiway.

All three aircraft and their passengers were bidden to a reception at BA's 'base' at the far end of the airport. In keeping with the mood, I opened the side window and stuck a small union flag on a stick into the crisp autumn air, waving it as I followed *Alpha Echo* and *Alpha Foxtrot* along the taxiway. They peeled off to do a celebratory lap on the airfield and to put us to the front of the line.

Then I saw what was coming: a water-arch reception from a Heathrow fire-tender. I whipped the flag in and tried to close the window, but the bloody thing had jammed.

Still fighting the damn thing – it had never given any trouble before – we rolled under the arch.

Perhaps sensing their opportunity, the crew of the tender tilted the waterjet towards our nose and what felt like the entire contents of its tanks gushed through onto the flight deck. Literally, fire hose volumes of water. How the instruments kept working I had no idea. I was soaked, but the show had to go on.

We parked up, awaiting *Alpha Golf*'s shiny white siblings and shut her engines down.

As I clambered out of my seat, one of the PRs in the back of the plane reminded me that my first duty, after clearing Customs, was to head for BA's posh, but temporary, three-storey media centre where I was required to do several more interviews for the evening news. I walked towards Richard Quest of CNN, whom I knew, and who was the first on the list. I'm naturally 'duck-footed', but today I really did waddle over, squelching like Daffy Duck and leaving a cartoon-like trail of footprints behind me. Richard was laughing so much we had to delay the interview.

As soon as the clutch of interviews was over, and I'd dried out a bit, I headed downstairs to the reception, which was taking place in our Concorde engineering hangar.

The drink and the conversation were already flowing as I walked in.

People were milling around a Concorde that had been polished to within an inch of her life.

As dusk fell, Rod Eddington, our CEO, and Sir Colin, our chairman, made speeches.

The exhilaration and excitement, the to-the-second timings, the party atmosphere on board the aircraft, all the messages of goodwill from air traffic and, above all, the sea of faces turned up at us as we'd swept in to land had combined now, with the help of a couple of glasses of champagne, to make me feel heady and slightly detached from the chatter and the chink of glasses.

The party continued well into the evening. At around 10:30, as the last celebrant headed out the door, I realised – still in my slightly detached state – that I was the last person still here. I picked up my cap and coat and headed out into the night air, where a car was due to pick me up and take me home.

There were few take-offs and landings at this time – Heathrow felt very still. A mist had descended. Under the sodium lights, five perfectly serviceable Concordes were parked up. The two other flyable aircraft in the fleet had been towed outside Engineering to stand beside the three that we'd landed earlier.

I stopped and stared at them. The mist gave the scene a slightly surreal air – as if I'd somehow wandered on to a giant movie set.

Five beautiful, perfectly serviceable aircraft stared back at me. They would never carry a fare-paying passenger again.

Why, I found myself asking, *Why are we doing this?*

It was as I turned away that the tears came.

42

The last flight of a Concorde in airline service wasn't quite the end of the Concorde flight story.

In the run up to the retirement, we'd discussed what we would do with the five flyable and three non-flyable airframes in our possession – two of the non-flyers being at Heathrow, with the third in a hangar at BAE Systems Filton, where it had been used for years as a spares repository. To be sure Concorde avoided being broken up – as had happened to most of BA's Tridents – we decided to get them as rapidly as we could to museums.

Which ones, though? We put out advertisements announcing our intention to divest the aircraft and invited proposals. We were overwhelmed.

In no time, we had received hundreds of responses, of which, we estimated, eighty were 'viable' – that is, they served what we felt were the purposes of Concorde's stakeholders – not least, the British public – as well as those of the petitioner.

Under the deal, the aircraft were offered at no charge, and we would deliver the flyable ones free too.

We then began our due diligence as to which of the eighty would be the most appropriate final resting places – an exercise we did in conjunction with Air France. Nowhere would get both a British Airways and an Air France Concorde, although quite a few places tried.

'Appropriate' to us meant ensuring they ended up in places

that had had a strong Concorde connection – as a destination or a manufacturing location.

New York City was a non-negotiable. We wanted to say thank you to this most hospitable of destinations by signalling we'd be gifting them an aircraft as soon as possible.

Barbados was another destination that had been intimately connected to Concorde and it also went to the top part of the list.

We had already decided with the French that the Smithsonian's Air and Space Museum in Washington D.C. should get one. The aircraft it received in the end coming from Air France.

As we at BA started to look around the UK, we took into account the fact that the Fleet Air Arm Museum at Yeovilton in Somerset already had a Concorde – the second prototype, 002 – and the Imperial War Museum at Duxford, Cambridgeshire, another: the first British pre-production prototype, 101.

As Filton had been inextricably linked to Concorde, Bristol was another non-negotiable – and given its significance to the aircraft's development and production, we felt it should receive a 'flyer' – not the one that had been stripped for spare parts in a hangar at Filton.

We decided that Manchester, which had been a great customer base for our charter flights and had shared its plans for a wonderful new 'Concorde Conference Centre' at Manchester International Airport, should also get one.

For many of the reasons we'd chosen Edinburgh as one of the start-points for the three-aircraft landing on the day of the last scheduled flight, we resolved it, too, should receive an aircraft – in addition to Edinburgh being a Concorde development and manufacturing hub, a 'Scottish Concorde' made sense for other reasons. In distributing airframes around the country, we wanted to ensure as many people in the UK as possible would be able to see them.

Having found homes for five of our eight Concordes, we turned our attention to the flight deck and cabin crew personnel who had served on her.

Concorde had been the last aircraft in BA service to have had a flight engineer. Under the terms of a long-standing deal with unions, 'The 1948 Agreement', British Airways had undertaken not to make anybody compulsorily redundant – a deal we'd stuck to; the flight engineers who'd flown on Concorde would be no exception. The options were two-fold. They could take early retirement, or they could be redeployed. Some of those that chose the latter path went on to become cabin crew, the deal being they would keep their engineer salaries and be privy to cabin crew allowances, which could be as much as £100/day.

Concorde pilots who were close to retirement could take early retirement. If not, then they could convert onto another type. Many chose the latter course, most redeploying on to short-haul rather than long-haul types because Concorde's rapid transit times – no matter the intercontinental distances – meant it had more in common with an A320 than a 747 in terms of flight times.

Because of the need to fly our five airworthy Concordes to their final resting places, I kept on a skeleton team of Concorde aircrew. Between five of us, we agreed we'd undertake the delivery flights to New York, Barbados, Bristol and Manchester. For logistical reasons, we subsequently decided that Edinburgh should have a non-flyer. We agreed in the end to dismantle *Alpha Alpha* and ship her from Heathrow to Scotland by road and by barge.

Where, then, should the final airworthy aircraft go?

We had received a very good proposal from the Museum of Flight in Seattle, Washington and decided this was where the fifth flyer should go. It hadn't been lost on us that Seattle was the geographical and spiritual home of Boeing. Half a century after the US plane-maker had bowed out of the supersonic transport game, Seattle would get a permanent reminder of what it had missed out on. To get *Alpha Golf* there we would need to fly her to New York, stay for a day or two, and from there head north and west along a supersonic corridor that paralleled the Arctic Circle. This, ultimately, is what happened, setting an airline east to west coast of the USA speed record that stands to this day.

This still left two non-flyers. *Alpha Bravo* was at Heathrow, where – with her galleys and toilets ignominiously removed – she had been used to trial the new cabin interior at the time of the retirement decision. As Heathrow had been *her* spiritual home, it was decided this is where she should stay.

To this day, that remains the case. For a year-and-a-half, rather mischievously, we parked *Alpha Bravo* right outside the Virgin Atlantic hangar, but she is now back where she belongs, outside our own, in a spot just north of the southern runway, where anyone landing from the east can see her out of the right side of their aircraft.

Alpha Bravo is still used as a testbed for BA's apprentices and engineering graduates. To ensure she doesn't tip on her tail as they tramp around her delicate insides, her nose has been weighted with ballast – thousands of old *High Life* in-flight magazines, initially, did the job. Occasional pressure from the British Airports Authority to move her has, thus far, come to naught. We always hoped the longer she stayed, the harder she would be to move. If she does get moved, we hope it will be to a place where the public can gain unfettered access to her.

Finally, this left *Delta Golf* – the aircraft that had languished as a spares base in a BAE Systems hangar at Filton.

One of the places that had pitched in with a request for an aircraft was the Brooklands Museum in Weybridge, just outside London. Brooklands had been the birthplace of UK motoring and aviation technology – something the museum celebrated wonderfully. The Vickers factory at Brooklands was also where a third of all Concordes had been designed and built before being shipped to Filton or Toulouse for final assembly. Also, there should be a Concorde available to the public near London, and Brooklands ticked that box too.

We came up with a plan, therefore, to ship *Delta Golf* from Bristol to Weybridge in bits – a journey, when it came to it, that saw her travel on the M4 and M25 to her final resting place. It was also agreed that Brooklands would act as a holding site for all the spare parts in our possession. This would ensure that Concordes at all the other museums would be supported technically for years to come.

There was, in addition, the matter of deciding what we should do with the aircraft's two simulators. Both – the simulator at Airbus Toulouse and the sim at Filton – had been put together using real Concorde nose sections. BA had become the owner of the Filton simulator following the 1984 restructuring deal, but we'd left it *in situ* because uprooting it at that time would have been too disruptive and risky.

Almost twenty years on, moving the sim still presented a challenge, but, if we wanted to preserve it for posterity, we knew we had little choice. In the intervening years, Filton had become a major wing-producing site for Airbus – there was no practical way of allowing the public on to the site.

Brooklands said it would take the sim – whether it would ever work again, however, was a different matter. In order to move it, all the wiring looms connecting the platform to the computers that had driven the visuals had to be severed. When the sim arrived in Weybridge with its wires hanging out the back, it looked like it would never work again.

In due course, Brooklands Museum established a team led by one of their volunteers, Gordon Roxburgh, whose 'day job' was Head of Production Technology at Sky Sports. The team included invaluable contributions from the University of Surrey, the Engineering and Physical Sciences Research Council and XPI Simulation Ltd. They not only restored it to full working order – making use of enhancements that have taken place in computer and projector technology since – but actually improved upon the original's visual reality.

Today, the sim is a major draw in its own right, its flight characteristics totally faithful to the real thing. Anyone, in fact, can now fly a Concorde, instructed by an ex-Concorde pilot, sometimes me. I love the looks on the faces of my 'students' at the end of a flight. Sheer amazement and pleasure – just like me when I did my first take-off and landing.

The aircraft itself, G-BBDG, has become 'The Concorde Experience' telling the story through displays, artefacts and a very realistic 'Virtual Flight', complete with motion simulation.

At the last count, three times as many people had flown the 'Virtual Flight' at Brooklands than had ever flown in any one individual BA Concorde during twenty-seven years of service.

The last ever flight of a Concorde took place on 26th November 2003 with the delivery of *Alpha Foxtrot* from Heathrow to Bristol's Filton Airport.

This would be an emotional journey in more ways than

one – not only were we taking Concorde back to her Bristol birthplace, but, thanks to special permission from the British Airways brass, we were bringing our families along for the ride. Thanks, too, to a desire to deliver the aircraft to Filton 'dry' – that is, after a supersonic flight to heat her up and evaporate any residual, internal moisture – we'd also wangled permission to take her past the Isles of Scilly, and 'round the bay', for one last Mach 2 supercruise around the block.

Because I'd piloted the final scheduled flight, I wanted to open this one up to the other pilots in my skeleton flight crew: Les Brodie and Paul Douglas.

'One of you flies captain, one flies co-pilot and I'll sit behind you,' I said to them. 'To decide who gets to fly, why don't you toss a coin?'

Which is what happened – and Les won. Accompanying us on the flight deck were two senior flight engineers: Warren Hazelby and Trevor Norcott.

After take-off from Heathrow, we headed out over the Bristol Channel exactly as if we were making for America. We accelerated to Mach 1, then Mach 2, shot up to 57,600 feet and hauled back around the Scillies for Filton. All in little over an hour. The plan called for Les to then fly in over the airfield at 3,000 feet, where a waiting helicopter would take some photos as she soared for the last time over the former BAC – now BAE – factory.

After this all went successfully and to plan, Les flew us back down the Severn Estuary, still at 3,000 feet, where further photo opportunities awaited.

'Look out for a couple of primary schools on your left-hand side, *Alpha Foxtrot*,' the controller said. 'There's something they want to show you . . .'

As I was in the jump-seat, it was easier for me to look out

the window than it was for Les or Paul. Just past Portishead, I spotted them, two groups of primary school children gathered on top of the cliffs, waving madly. As we roared past, each child lifted a piece of white card, which formed into a familiar shape. Seen from the height we were at: a perfect silhouette of Concorde.

The route back to Filton took us over Weston-super-Mare and Bristol Lulsgate Airport before passing the Clifton Suspension Bridge to the west of the city.

Unlike our sojourn along the Severn, this part of the day had been scripted in meticulous detail – to the extent that Les, Paul and I had come down to Bristol the week before to fly in the helicopter photo-ship that would rendezvous with us over the bridge. The purpose of all the planning was for the photographers to get what we hoped would be a memorable last shot – one Bristol-born icon set against another: Brunel's Victorian masterpiece.

To execute the picture, we needed to be mindful of the angle the helicopter would be shooting from. To make it look like we were over the bridge, we worked out that we had to be positioned over the Regency Royal York Crescent that sits near the top of the Clifton cliffs. As we flew back up the Severn towards the bridge, we suddenly became aware of the crowds that had gathered on the clifftops. We'd been extremely lucky with the weather. A thunderstorm had just rolled through, and another was due in an hour. The people below had braved the elements in their thousands to come and see us for the last time. But the miracle was that gap in the clouds – a clear November sky into which we soared for the last time for the loyal crowd that had turned out to see us.

After the helicopter crew had grabbed its shots – photos, I reckoned later, that were amongst the very best ever taken of Concorde – we waggled our wings and set course for Filton, where Les greased the tarmac in an immaculate landing and we shut down our four Olympuses. After a whirl of ceremonial theatre that saw us hand *Alpha Fox*'s maintenance log over to the authorities, it suddenly hit me: no more delaying of the inevitable; this was it.

After all the dignitaries had gone home, somebody asked what we were all going to do. None of us had thought as far ahead as the evening. We were pretty much washed out. The last thing we wanted to do after a day like today was go somewhere where people wanted us to retell all the old stories.

Paul Douglas, who had been today's co-pilot, leapt to the rescue. He and his wife invited those of us who'd flown the plane – accompanied by our wives, who'd come along, too – back to their very nice house near Bath. 'Come have dinner, stay the night, and shoot off back to London in the morning,' he said.

Which is what we did. And it was perfect. In the company of colleagues, we were under no pressure to say or do anything. Just 'shoot the breeze'.

And because there was no pressure on us to tell any stories, we ended up, of course, reliving them all. We told and retold all the best ones into the night like there was no tomorrow, knowing that this really was it; there *was* no tomorrow – not for *our* aeroplane, anyhow. We laughed and we cried – and, at some point, fortified by Paul's finest malt, I uttered a quiet prayer of thanks.

To Concorde and all the people who had made her possible:
Thank you for an incredible twenty-seven years of service.
Thank you for twenty-two of the most magical years for me.
And for making the dream of this seven-year-old boy come true.

I knew on the day I joined BOAC, on the 27th May 1969, that the date of my retirement would be the 11th May 2004. Soon after I'd flown Concorde for the very last time, it was inevitable, I supposed, that my thoughts began to fix on this date. I was fifty-four years old and pilots at that time with BA had to retire at fifty-five. So, what next, I asked myself? One of my options was to do a conversion course on one of the aircraft, as general manager, that I was responsible for: the A320. But my boss had a different idea. Would I take on the role of General Manager at Gatwick Airport in addition to my Heathrow medium and short haul responsibilities? I enjoyed a challenge and one office at Gatwick, another at Heathrow, with three days a week at each, swapping Concorde flying for time on the M25 and the M23 in my Ford Galaxy, was, well, a new kind of challenge.

Why not?

The only real question it raised was whether I should swap my pilot's contract for a management contract 'with a supplement to fly' or keep my pilot's contract with a 'supplement to manage'? In terms of take-home pay and pension rights, it didn't make much difference, but it might make a big difference at the next 'night of the long knives', when BA came to culling managers, a not infrequent exercise. I might then find myself out on my ear.

When I rang BA's pension people for advice, a dispassionate male voice at the other end of the phone said: 'We can't advise you, I'm afraid.'

How bloody helpful, I thought.

'I'm sorry, but they're the rules,' the voice said, flatly.

I was about to hang up, when the line unexpectedly crackled back into life: 'What I can provide you with, however, are some statistics.'

'Statistics?'

'Yes. Statistically, of 100 per cent of people in your position, 98.5 per cent choose to remain on a pilot's contract. That's all we can say on the matter.'

It was enough. If 98.5 per cent of people in my position chose to remain on a pilot's contract, there had to be a clue in there somewhere.

My decision had been made for me: I'd take the job, but I'd stay on my pilot's contract and retire at fifty-five: the 11th May next year would be the date. Chris and I had always had a life plan that had included this moment. Amy would be ten by then and we'd agreed we wanted to invest time in seeing her grow through her teenage years, to travel, and generally take life a bit easier.

A couple of weeks into the new job, I was asked if I'd like to attend a ceremony at the giant Airbus facility in Toulouse that would see the European aerospace consortium surrender Concorde's Type Certificate and Certificate of Airworthiness back to the regulatory authorities. I, of course, said yes.

The date chosen was an auspicious one: 17th December, exactly one hundred years from the date that the Wright Brothers had achieved sustained powered flight. The ceremony would be low-key, but poignant. British Airways would be represented. Air France too. All the manufacturers and regulators.

As I stepped on to the Toulouse tarmac on what was – for southern France – a cold December morning, my mind drifted back to the moment earlier in the year when we'd been approached by the US authorities tasked with commemorating the seminal Wright Brothers moment. They had wanted to know if BA would contribute a Concorde to the fly-past planned above the dunes at Kill Devil Hills, near Kitty Hawk, North Carolina, at the precise hour that the Wrights' little Flyer had made its historic twelve-second flight – 10:35 a.m. I'd thought it was a terrific idea but was told that hanging on for another month beyond the delivery of *Alpha Foxtrot* to her last resting place was, from a cost perspective, unsustainable. And, needless to say, they were right.

Even so, it was hard, as the paperwork was handed over – the mood amongst all of us present as sombre as if we were at a church service – for me not to think of the preparations under way several thousand miles away on the US Eastern Seaboard; for Concorde, just three-and-a-half hours' flight time distant. But the weather gods had the last word. It turned out that, on the morning of December 17th, the weather around Kitty Hawk was appalling. Such low cloud and visibility that even Orville and Wilbur – stoics that they were – would have cancelled.

As our aircraft taxied out to the threshold on its way back to Heathrow, we passed by the site of the former Concorde support facility that I knew so well and was shocked to see that it had already gone. Contractors in hard hats and high-viz jackets were swarming over the foundations of what would soon be a vast new building – the support facility for the new A380 super-jumbo.

Half an hour into the flight home, I told myself I was ready for retirement; that it was time for a different kind of fun. Across the choppy waters of the Channel, however, my old friend *Madame La France* had other ideas.

43

I had been retired nearly two years – and enjoying every moment of it – when I got a call out of the blue from an American lawyer, who was contacting me as result of a conversation that had taken place between – I heard him shuffling his notes on the other end of the phone – 'a Leigh Skinner', whom he said was known to me, and a senior exec from Continental Airlines.

I did indeed know Leigh, who was the inflight international service director for Continental Airlines and a friend of Chris's of very long standing.

The legal eagle explained that Leigh was fed up with all the bad press her company had been getting about the accident and knew that Continental was likely to be blamed.

So, she contacted a very senior member of their 'Inflight Service' department and said: 'I know this British guy who knows a lot about Concorde and might be able to help us.' She went on to explain my connection, my twenty-two-year knowledge of the intricacies of the aircraft and my belief that no accident was 'single-cause'.

The exec confided in Leigh that they were in a bit of a bind and, indeed, needed some expert advice on the workings of the recently retired, world's only supersonic passenger jet. Next the exec contacted their legal department, who in turn spoke to retained lawyers, and the first thing that I knew about all this was a call out of the blue from a US law firm based in Chicago.

The lawyer explained that his job, and that of his team, had been to assess a technical report that had recently been published by the French prefect and his team of investigators.

The report was widely viewed as the opening shot of what might likely soon be a full-blown prosecution case. The advice assembled by the prefect in the report had been prepared in conjunction with a team of aviation specialists, the lawyer told me, but had not, in so far as anyone on the Continental side could tell, included advice from any *Concorde* specialists.

The lawyer cleared his throat before coming to the point: 'Would you be able to consult to us on the veracity and accuracy of the advice that is within the report as well as on the conclusions drawn, Captain Bannister?'

'Please, it's Mike,' I told him. I asked him to send me what he'd got so I could make a decision.

After I'd studied it, I called him back and said that I would.

The purpose of the report, I learnt when the first chunk arrived by courier, was to assist the prefect in determining whether a prosecution was appropriate and/or likely to succeed. It went into enormous technical detail about Concorde and its technology. Some of the conclusions the prefect's experts had drawn were correct, some were incomplete, some were off and some plain wrong.

My job wasn't to comment on the stuff that was right, but on any material of interest. Sections would be sent to me for my comments, I'd go through them, and send my notes back.

This process went on for the best part of a year.

Things went quiet for a while, but then I was contacted by another lawyer who was a part of Continental's main legal team in Houston. Would I be interested in being a technical adviser, they wanted to know, in any case that might be brought against the airline?

When I said that potentially I would, they suggested that I came over to Houston for an initial meeting.

I met and talked with the airline's senior lawyer and a lawyer representing their insurers. We agreed that the best thing I could do was to act as the layperson's guide to the complexities of Concorde, given that the technical aspects of Concorde would be central to any case.

The kinds of areas that would need to be decoded were all the areas that had been highlighted by what I (and the UK air crash investigation authorities) had considered to be a part of the flawed, French-driven accident investigation and the report that had been published in 2001: the nuances of refuelling; the aircraft's all-important centre of gravity; operating techniques, particularly prior to take-off; and human factors in terms of the actions of the flight crew and the comments they had made on the flight deck – comments that had been recorded, of course, by the cockpit voice recorder. In that initial meeting with Continental's lawyers, I'd talked generally about my feelings on the report – the whole notion that accidents weren't 'single cause events'; that there were always a set of links in what was known as a 'causal chain', some of them major, some minor, but all highly germane to what had happened.

When the discussion turned to the actions of the crew that day, however, I began to feel uncomfortable; and told them so. It's a sort of unwritten rule you don't criticise other crews; that it was all very well being wise after the event, but who was to say *you'd* have done any differently?

The crew in this case were dead, unable to defend themselves or give an account of their actions.

'Look,' I said, 'I'm perfectly happy to help you, but if this goes any further, I'm not prepared to stand up and criticise Captain Marty and his crew.'

OK, they said. We get it.

And for the moment, that was that. But back in the UK, the thought gnawed at me. What was the right thing to do? On the one hand, there was the code – the unwritten rule. On the other, the crash at Gonesse and its aftermath – the shocking nature of the BEA report – had felt like unfinished business and I wanted to see justice done; and, as pertinently, I felt, a need to make good on Amy's wish: that nothing like this should ever happen again.

The thought came to me that I could evaluate the actions of Marty and his crew that day as I would have a living crew – as if I were the check pilot, observing them had I been their examiner; as I would have were I back in my role as a training captain and the crew of Flight 4590 were being put through a routine route-check.

Had I had that role on that day, what would I have said to them?

These wouldn't be comments to a third party. I imagined myself speaking directly to the crew themselves. Could I do that, I asked myself.

I knew in my heart after a sleepless night that this was a way – the only way, in fact – for me to be able to do the right thing: getting to the bottom of the events that had led to a crash that had killed 113 people; seeing justice done; and ensuring – in so far as I could – that it never happened again.

But there was another reason, too. It was clear from the way the prefect's report had focused on certain things that a light was being shone in a particular direction.

This became clearer still when the French authorities fired

their first salvoes in what would become almost seven years of involvement for me in the prosecution of those they accused of being responsible for the crash.

Those salvoes came with the announcement on 3rd July 2008 from the prosecutor's office in Pontoise, a suburb of Paris, that five people would stand trial: the director of Aérospatiale's Concorde programme from 1978 to 1994; a former chief engineer of Aérospatiale's engineering team; an ex-director of France's civil aviation authority; two Continental Airlines employees; and Continental itself.

The director of Aérospatiale's Concorde programme at that time had been *Le Père du Concorde*: Henri Perrier.

Henri, along with two more of the five, the prosecutor asserted, had been fundamentally associated with the design and certification of the aeroplane. They were 'legacy post holders', not in place when the original design decisions had been taken, but aware, allegedly, of the risks that the aeroplane had represented – and supposedly had done nothing about them.

The other two had been with an airline – an American airline, at that – unconnected to the crash except for what the prosecutor alleged had been its poor maintenance procedures, which had led to a piece of one of its aircraft – the DC-10 – being on the runway that day.

Nowhere had a light been shone on Air France, despite what investigators – on the UK side of the Channel, at least – had identified as distinct irregularities in its procedures.

We knew, for example, that the missing copies of the all-important load sheet for Flight 4590 had never reappeared – the document that would have shown where all the baggage had been placed; how much fuel had been loaded – and where; what the aircraft's vital centre of gravity had been; and all the ramifications of those issues, including instructions to the crew about how much fuel should have been burnt prior to take-off. A copy is always given to the captain and a copy is always retained at the airport you depart from for a period of time after the departure. Those are the regulations. No ifs; no buts.

The second irregularity surrounding the paperwork (it was, in actual fact, of course, computerised) for AF4590 that day concerned the fuel loading sheet – that is, fuel advice to the crew on what fuel has been loaded and where – which looked as if it had been tampered with.

A third irregularity was equally baffling – there was absolutely no sign of any internal Air France investigation report. Surely, there must have been one?

Had it gone missing as well?

Because the light had been directed away from Air France, it was self-evident that it had to illuminate something else. The five accused were, to my mind, the unfortunate souls on whom this spotlight had landed – and they would soon stand trial for their supposed sins.

But the unnamed co-accused in the case, judging by where the light from the prosecutor's office was shining brightest, was Concorde herself.

I thought back to those three days of sunshine at Istres, as Henri and I had leaned against the wall of the Portakabin – he in his old hat, me in mine.

He had known then – it seemed to me he had known all along – that this day would come.

Any reservations I'd had about doing the right thing were progressively swept aside the more I got into the data. The aeroplane that had given me twenty-two of my best career years, together with those responsible for it, was under attack.

People were being blamed for what appeared to be a string of institutional failures; failures that needed to be brought into the light of accountability were an event like this to be prevented from happening again.

It wasn't just in the missing paperwork or the fact the causal chain had been replaced by what had always seemed to me to be a spurious point failure for the events that had unfolded on 25th July 2000: that piece of metal.

Other elements had been glossed over like they had never existed.

For instance, it became clear when we'd attempted to examine the parts that had been brought into the hangar at Le Bourget that things were amiss – literally as well as metaphorically. After we'd developed the hand signals that had allowed us – the British investigative team – to zone in on items of interest without the French noticing what had grabbed our attention, a senior UK engineer had called me over to a particular collection of debris.

Like something out of a bad movie, my memory told me he'd dropped to the ground to tie his shoelaces next to the part he'd wanted to draw my attention to – a section of the left-hand undercarriage assembly – but this might have been the way I've come to see it in my mind's eye since. In any event, he'd whispered to me: 'Mike, look down there. What do you see?'

I peered closer, hoping the attention of the *gendarme* tasked with watching over us as well as the wreckage had been distracted elsewhere.

'The spacer's missing,' the engineer hissed before I could reply.

The spacer was a cylindrical tube, about twelve inches long and five inches wide, which sat inside the undercarriage. Its job was to keep the two sets of wheels apart – to stop them acting independently, like a wonky supermarket trolley.

Subsequent examination of the wreckage showed that an Air France engineering team that had carried out some routine maintenance on AF4590 on 17th and 18th July – a week before the crash – had omitted to replace the spacer on the left undercarriage bogie. Eyewitness testimony, evidence from the cockpit voice and crash data recorders, as well as tyre marks on the runway, had shown unequivocally what crash experts had known pretty much from the start: that Captain Marty had rotated – that is, he had pulled back on the stick to lift Flight 4590 into the air – a full 15 knots before he should have.

This was very strange because Concorde pilots had been taught that 'speed is your friend'; the aircraft performed better the faster she went. Never get slow or do anything at too low a speed, especially on rotation, where, in the event of an engine failure, accuracy was really important in getting to the right climb-out speed: V_2. Performance deteriorated rapidly if you didn't reach it.

No definitive explanation for Marty's decision to do what he'd done had come to light then – or in the years since.

Almost a year after the crash, however, there had been a flurry of speculation in the UK media – some of it very well informed – that the omission of the spacer had been a vital link in the causal chain leading to the crash.

The BEA's accident report had concluded that 'nothing in the research undertaken (indicated) that the absence of the spacer contributed in any way to the accident on 25 July 2000.'

Others, though, disagreed. A well-researched article that had appeared in the *Observer* newspaper on 13th May 2001, almost a year after the crash, questioned the whole basis of the French authorities' refusal to give the causal chain argument – the start-point of almost all credible air accident investigations – due credence. The spacer issue seemed to exemplify this. As the newspaper pointed out, the Air France maintenance team had replaced a 'lifed' undercarriage beam, a horizontal tube through which the two wheel axles passed at each end. In the middle of the beam was a pivot joint that connected the beam to the vertical leg of the landing gear. The key parts of the pivot that bore the load of the aircraft during take-off and landing were two steel 'shear bushes'. The foot-long spacer was designed to keep the bushes in position, but, following the maintenance procedure, AF4590 left the hangar without it having been replaced – after the crash, investigators found it in the Air France workshop still attached to the old beam, which had been replaced.

Examination of the wreckage showed that there were so-called 'witness marks' inside the undercarriage – four clear physical marks that verified the aircraft had done two take-offs and two landings between the maintenance procedure and the crash – which, indeed, it had; it had been to New York and back twice. The marks showed that with each take-off and landing the bushes had wandered further and further apart. By the day of the crash, they had moved a full seven inches out of true. Instead of being held firmly in a snug-fitting pivot, the beam and the wheels were wobbling, like that wonky supermarket trolley, with three degrees of possible motion, left or right.

As AF4590 taxied towards the runway threshold on that fateful 25th July, there was nothing, so this theory went, to ensure that the front wheels of the left-hand undercarriage continued to remain in line with the back wheels.

A veteran Air France Concorde captain, Jean-Marie Chauve, and a senior flight engineer, Michel Suaud, had recently spent six months preparing an independent sixty-page report on the accident. Their conclusion was that the left set of wheels was already out of alignment when AF4590 started to roll down the runway.

Both noted that the aircraft's acceleration on take-off had been slower than usual – that 'there was something retarding the aircraft, slowing it back'. This, their report alleged, could only have been friction from the left set of wheels, something the BEA report denied; it said the acceleration had been normal until the tyre burst – a failure it had pinned squarely on the wear-strip that had been dumped on the runway by the departing Continental DC-10.

Chauve and Suard went on to allege that this friction, driven by the massive thrust of Concorde's engines, had prompted the front inboard tyre to burst, causing the two front wheels to hard-over to the left. The 'smoking gun', they said, was a remarkable series of photographs in the BEA's own report that showed 'unmistakably the skid marks of four tyres heading off the runway on to the concrete shoulder, almost reaching the rough grass beyond'.

The BEA report did indeed depict the aircraft's tyre marks veering markedly to the left, but the French accident investigation authority, for some reason, seemed keen to avoid the term 'skid' as a description of what had happened in the run-up to Marty's decision – for whatever reason – to rotate the aircraft a full 15 knots too soon. It maintained that the leftward 'yaw' was caused not by the faulty landing gear, but by the loss of thrust from Engines 1 and 2, the two engines on the port-wing, caused by a 'surge' – an increase in rpm due to the fact that the engine's compressor blades had stalled. This usually happens when airflow entering the engine does so at the wrong angle, typically at high angles of attack. The BEA report had failed to determine why Engines 1 and 2 had suffered a loss of thrust, there being evidence, it stated, of both foreign object damage, 'probably linked to the explosion of the tyre,' and the ingestion of hot gases, 'probably caused by the ingestion of a kerosene/hot gas mixture, facilitated by the change in the aircraft's attitude'.

Be that as it may, the fact that the BEA was attributing AF4590's 'yaw', 'skid' – call it whatever you like, because to me these were just semantics – to a loss of engine thrust was, to my mind, disingenuous in the extreme.

For a start, the BEA's own report stated that thrust from Engine 1 was very quickly restored.

Second, as any Concorde pilot knew from umpteen engine-out simulations, the aircraft's engines were positioned much closer to the centreline than a conventional aircraft (like a Boeing 747) and it was relatively impervious to marked yaw moments during engine-outs. I'd overseen Concorde captains and first officers in the simulator experience the loss of an engine at or around V_1, get airborne and climb away safely, no problem.

What this all added up to, for me, was a pattern of disregard by the French authorities for evidence that should have been examined thoroughly – not dismissed in the way it had been in the report and by the prosecutors who'd brought the case.

As I mulled the right thing to do, something else bubbled up.

Pursuing Henri was, for me, further evidence the French had got this all wrong.

I called the lawyers acting for Continental and told them I'd do it.

44

In February 2010, the trial of the five defendants and of Continental began.

In addition to Henri, the others were: Jacques Hérubel, also of Aérospatiale; Claude Frantzen, head of the DGAC, the French equivalent of the UK Civil Aviation Authority; John Taylor, the Continental mechanic who replaced the wear-strip on the DC-10; and Stanley Ford, Taylor's manager. Ford had been charged with negligence for allowing the repair to have gone ahead. The three Frenchmen had been put in the dock for allegedly allowing Concorde to fly knowing that its fuel tanks could be susceptible to damage.

Absent from the charge sheet, but on it, nonetheless, was Concorde herself.

Soon after the trial opened at the Pointoise Criminal Court, I headed over to Paris to meet with the French legal firm that had been hired by Continental to act as its in-country defence team. Metzner and Associates was headed by Olivier Metzner, a top-notch, larger-than-life character, who came across on our first meeting – and subsequently – like a kind of French Perry Mason.

Olivier had just cemented his reputation as being amongst the best of the best criminal lawyers in France. Weeks before, he had secured a win – an acquittal on all the charges – for his client Dominique de Villepin, ex-Prime Minister of France, who had been accused by Nicolas Sarkozy, the President of France, of complicity in allowing false accusations of corruption against Sarkozy to proceed and circulate. Amongst others he'd represented were Manuel Noriega, the Panamanian dictator, the Church of Scientology, and a group of French citizens jailed in the Dominican Republic for cocaine trafficking. It worried me a little that Concorde's fate had been mixed with such company, but common sense said I had to let Olivier's record speak for itself.

He was as impressive looking as his reputation suggested: grey-haired, wearing his glasses on the end of his nose and rarely without a cigar in his hand.

On our first meeting, his distinguished appearance made me think he was years older than me, when, in fact, we were almost the same age. He hardly ever spoke English, although it was evident from the way he listened to me and the other English-speakers on the Continental team – and occasionally interjected – that his command of English was very good.

He didn't participate directly in the discussions that took place in his palatial offices near the Musée d'Orsay, but he was usually there, in the background, pacing the room, listening.

In hindsight, I could see he was checking me out to see if I would be able to handle being a witness. They had already retained one expert witness, Gary Wagner, an ex-Air Canada pilot and a veteran of aircraft crash investigations. Gary was exceptionally good at what he did and had served frequently as an expert witness during events such as these, but, crucially, he didn't have Concorde experience.

For several days, as we went over the pivotal points, I could feel Olivier's eyes boring into the back of my head from the back of the room.

I was used to standing in front of groups of people, trying to explain in layman's terms the complexities of the aircraft

I'd flown – the fact I'd been an instructor helped too. But I had never been an expert witness in a criminal trial before and the way that the trial was conducted – especially in France – was, thus, completely unknown.

What I appreciated very much, therefore, was the patience with which Metzner and his team briefed me on what I might expect when I entered the courtroom.

'There's a panel of judges, three judges – well, actually, four judges,' a junior member of the team explained to me when it became clear that they would put me forward. 'The fourth judge is, like, how do you say, a *stand-in*?'

'A substitute,' I suggested. 'In case of sickness or something?'

'*C'est ça*,' the young lawyer said enthusiastically. 'Of the three judges who preside over the case, one of them acts as *le président*, the one who sits in the middle. All the questions are, in theory, directed through this person. The trial, *naturellement*, is conducted *en français*. Do you speak French?'

'Some,' I replied.

'*Bien*,' he said. 'Then probably best to keep this to yourself. There will be a simultaneous translator, but you will need to speak slowly so that the translation can keep pace with your testimony.' He smiled and added: 'If you understand more than the prosecution thinks you understand, then this is maybe a good thing, yes?'

'Yes,' I smiled back. Knots, though, began to form in the pit of my stomach just thinking about it.

When it came to the day, the lawyer said, I would take the stand in front of the bench of judges and would be asked a series of questions.

'You will be allowed to give your *témoignage* – your . . . testimony – and your assertions, and then you will be questioned on them. The defence lawyers – us – will be sitting on the left and the prosecution's lawyers on the right.

'You must understand, *Capitaine* Bannister, that the system in France is different than the system in the UK or in America. There is a strange relationship between the chief prosecutor and the president of the court. *En effet*, the chief prosecutor is acting on behalf of the French state, a curiosity in the legal process we inherited from the era of *Napoléon*. And this curiosity is that you are, *en effet*, presumed to be guilty until proven innocent – and the guilty . . .' he stopped to put quote marks in the air with his fingers around the word 'are on trial because the prosecutor has said so. This process, then, is about trying to establish whether the prosecutor's assessment is correct. *C'est clair?*'

I nodded. *Clear enough*. I mightn't have liked it, but this was how it worked here.

There was one other thing I needed to be briefed on before my appearance before the judges, the lawyer said. I would have no knowledge of what had been imparted in the courtroom prior to my testimony. I couldn't be told what had been said other than that which had been reported publicly.

On the day, when I arrived in court, I was further advised, I would be kept isolated in a room away from the trial itself, until such time as I was summoned. If there was anybody else in this ante-room waiting to give testimony, I was under strict instruction not to communicate in any way.

Once I had given my testimony, only then would I be able to sit in the courtroom with Metzner and his team, dispensing my advice as needed.

Olivier knew Henri Perrier and I were friends. I asked his

colleague, therefore, whether it would be possible to speak with Henri in or outside the court.

'Although technically we are all on the same side, Monsieur Perrier is in his own battle with the prosecutor.' It would, therefore, he said, be inadvisable to have any contact with him or any other of the defendants for the duration of the trial. The case, he added, was on a knife-edge as it was.

'What are our chances?'

'Of success?'

I nodded.

'At this point, I think we would really prefer not to say.'

As we were introduced to more and more evidence, Gary Wagner and I came to believe that Continental's chances of acquittal were around sixty-forty. But our French colleagues remained far more sanguine. It emerged through these and other discussions with Metzner and his team that, in their eyes, the chances of the verdict going in our favour were less than fifty-fifty.

45

It was the summer before I got my day in court.

Our taxi pulled up in front of the building, which was not what I expected. In place of a town hall affair with steps leading up to big wooden doors between classical pillars, the Pontoise High Court was a tall, modern building perched on a busy corner with architectural influences, it seemed, from London's Royal Festival Hall: all concrete and glass under a flat roof.

As Olivier and I tumbled out of the taxi, a handful of reporters rushed forward. Olivier brushed them aside and ushered me up the steps. Inside the building, he left me to check in with the clerk of the court with help from one of the Continental lawyers, a lady called Carson Seeligson. Without even looking up, the clerk, a small man with a pinched face, asked me to hand over my passport.

As ID checks went, this wasn't so unreasonable, but nobody had told me there would be a requirement to prove I was who I said I was. I'd been so wrapped up in the testimony I was going to give that the mere *bagatelle* of my actual identity had never entered my head. 'I'm sorry,' I said, 'I don't have any.'

'*Comment? Quoi?*'

'I don't have any,' I said again. 'Not with me.'

'This is Captain Michael Bannister,' Carson explained to the clerk. 'He is giving his testimony to the court today.'

With a Gallic shrug, the clerk pointed to the sky, as if to

say, I don't care if he's God Almighty, he doesn't come into my court without proper ID.

Via a series of hand-gestures, we were able to communicate that my passport – the only *carte d'identité* I had with me – was in my hotel room in central Paris – a good hour's drive away – and with the need to deliver my testimony imminent, impossible, therefore, to collect and bring back in time.

Do you have any other form of ID with you?

It was then Carson hit on what I considered to be a very improbable solution: why didn't he go ahead and just Google *Le Capitaine Bannister*?

The clerk disappeared – to make a complaint, to summon an escort to eject me from the building, I had no idea – only to reappear moments later with something approaching a smile on his face.

Google had delivered the goods: I was who I said I was. '*S'il vous plaît*,' he said, gesturing towards the metal detector and the interior of the building. '*Passez.*'

I was ushered into the isolation room – a small, windowless, featureless space with a couple of metal chairs. Mindful of the directive – that what I read in here couldn't in any way be associated with the trial – and feeling that cameras were monitoring my every move, I reached into my briefcase and pulled out my book – realising, only as I did so, it was a John bloody Grisham.

One, two, three hours dragged by. Then an official walked in to tell me that the proceedings had overrun, and that I wouldn't be required that day.

'When, then?' I asked.

The official wasn't sure.

I returned to the hotel with Olivier and his team, the bus weirdly silent because no one could communicate in my presence anything about the day's events; an official silence that continued to hang over our dinner that evening.

We assembled the next morning at Olivier's offices for coffee and croissants before driving out to Pontoise. I sipped my coffee but was far too keyed up to eat. Although it was against the rules for the legal team specifically to coach me, they had let me know I could expect some intense interrogation during the day's proceedings about the status of Flight 4590 on the day of the crash.

Back at court, this time with my passport, I sat in the bland, bare ante-chamber again, until summoned by an official to make my appearance.

I entered the court via a side door. Like the exterior, it was modern and functional, the windows extending from head height all the way to the ceiling.

Benches, arranged like the pews of a church, ran from front to back on both sides, the dark, polished wood offset by a red carpet, adding to its High-Church feel – quite an irony, a small voice told me, in a country that was so proudly vocal about having cut all those ties between church and state.

From the front bench, the four female judges – dressed in black with white court bibs hanging from their throats – regarded me stonily.

In front of them was a glass cabinet containing items of evidence.

Amongst the artefacts on show, I spotted the titanium wear-strip that had been so elusive in our investigation and a large piece of AF4590's tyre.

Beneath the cabinet was a small podium, to which I was now directed.

Behind me, benches were packed with the relatives of those killed.

As I looked up, I spotted Olivier fiddling with his *jabot* – the white ruff around his neck. He caught my eye and gave me an encouraging wink. Next to him were various members of his team, including Continental's US lawyers.

Across from the defence were the prosecuting attorneys. Between the two sides, ranged along a bench facing the judges, sat the five defendants. The closest, one of the Americans, was no more than a few feet away.

As I walked up the steps to the witness stand, one of the defendants turned. It was Henri, who somehow managed to give me one of his broad smiles and a thumbs-up.

I gave him a faltering smile back.

A court official stepped forward to help me with my earpiece. Then, an instruction to repeat, after him: '*Je jure de parler, sans haine et sans crainte, de dire toute la vérité, rien que la vérité . . .*'

'*I swear to speak, without hatred and without fear, the whole truth, nothing but the truth . . .*'

As I had been briefed, the translation began to come through the earpiece and was a few seconds behind what was being said in the room.

I turned to the bench. The presiding judge, Madame La Présidente, continued to regard me without a flicker of emotion. She was, I guessed, in her mid-fifties, with mousey shoulder-length hair. Her glasses had thick brown frames. She began with a summary of my experience – twenty-two years as a Concorde pilot; and, in amongst that experience, time spent as the chief pilot, the chief training manager, the chief technical pilot and, lastly, type test pilot.

'We would like to begin by speaking with you today about

your impressions of the crew of Air France Flight 4590,' she then said.

Olivier was already on his feet. He and the judge exchanged a few words, which weren't translated, but which I understood: 'Your witness, Maître Metzner . . .'

The thing I had been so reluctant to do – to criticise the actions of the crew – had reared its ugly head from the start. I reminded myself about the promise I'd made – that the only way I could do justice to the situation was by imagining myself as the check pilot on the flight deck – watching over Captain Marty, First Officer Marcot and the flight engineer, Jardinaud, on that 25th July.

'How would you characterise the flight on this day – the day of the accident?' Olivier asked, reminding me to direct my response to the judge.

'Everybody would have been under intense pressure,' I replied.

I pointed to the fact that the aeroplane had been delayed by two hours but that it had to be in New York on time so the passengers could catch the ship.

When I finished, Olivier said: 'You have a particular theory, I believe.'

Suitably cued, I outlined the theory that Tank 5 had been 100 per cent full because of the unapproved fuelling procedure and that switch positions found at the crash site appeared to confirm this. I produced a lock-toggle switch for a 'show and tell' and a two-foot model of an Air France Concorde marked up in multicoloured ink to show the layout of the tanks and pumps.

I pointed out certain other actions that had been irregular that day: the 60 kilograms of newspapers bundled on board that hadn't been logged; the bags thrown into the rear

baggage hold; the lack of any formal – or *informal* – acknowledgement in the cockpit regarding these discrepancies; the
lack of the key pieces of paperwork that would have told this
story . . .

And others – the fact that the fly-by-wire system had
flipped from its Blue to Green standby mode without any of
the crew calling for the checklist or the minimum dispatch
list. And, finally, the fact Marty had taken off having received
notification from the tower that there was a tailwind on Runway 26R – knowing he was a tonne-and-a-half over maximum
take-off weight and 6 to 8 tonnes over the aircraft's all-critical
performance-*limited* take-off weight.

When I finished, Olivier paused. He studied his hands,
turning them over several times. 'What would you have done
under those circumstances, I wonder?' he asked finally.

I thought about my words carefully. 'If I've planned my
take-off under one set of conditions and those conditions
significantly change, as happened with the change in wind
direction, the very least I'm going to do is discuss it.'

'What would you say and do as a check-pilot if a crew you
were evaluating took an action such as that taken by the crew
of *Capitaine* Marty?'

'I would have been very critical,' I said.

Oliver looked at me over his glasses. 'It would be useful
for the court to know whether in your role as a senior Concorde instructor you have ever had to – I think the word is
"intervene" – for a very serious offence such as this.'

'Only once.'

'And what happened to this individual?'

'We had to let him go.'

'You mean, he lost his job?'

'Yes.'

'I see,' Olivier said, glancing at the judge. 'What options did the crew of Air France Flight 4590 have at the point that the wind changed?'

I ran through them. They could have waited for it to drop. They could have waited to burn off the requisite amount of fuel. They could have taxied back to the terminal to offload some fuel or bags or some passengers, even. They could have taxied to the other end of 26R to take off into the wind. 'Or . . .' I stopped.

'Yes?' Olivier said.

'There was supposed to be a process. The process is that you're supposed to discuss the actions you take, especially if the operating conditions change in some way. But they didn't. There *was* no process.'

'That is a serious charge, *Capitaine* Bannister.'

'Yes,' I said. 'And it saddens me deeply to make it.'

I raised my eyes to the judges to see if any of my arguments had registered. But the look I got back was the same I'd got when I walked in.

I felt like kicking myself. What had I been thinking? That I'd walk into the court, outline the whole causal chain argument and the prosecutor and the judges would look up and say: *Oh my God, we'd never thought of that. Thank you, Capitaine Bannister, for clarifying this for us. Now, we can all go home.*

After my testimony, Olivier and his team were kind enough to inform me I'd done well, but the looks on their faces told me what I didn't want to know.

We were going to lose – the aeroplane would be found guilty.

Worse, innocent people – including Henri – were going to go to jail.

46

It took another six months for the verdict to come in.

On 6th December 2010, the judges found Continental guilty of criminal responsibility for the crash. It was fined 200,000 euros and ordered to pay one million euros to Air France. Also found guilty was John Taylor, the Continental mechanic who had allegedly been negligent in fitting the wear-strip to the DC-10. Taylor was fined 2,000 euros and given a fifteen-month suspended prison sentence.

Acquitted that day were Stanley Ford, Taylor's boss; Jacques Hérubel of Aérospatiale; Claude Frantzen, the former head of the DGAC; and, to my great relief, Henri.

I was at home when they announced the verdict. It didn't take long for the phone to ring – a call from a lawyer at Metzner's. The verdict was wrong, he told me, and Continental would be appealing. Was I up for helping them again?

As I understood the appeal process in France, the prosecutor also had the right to appeal against the non-guilty verdicts brought against Ford, Hérubel, Frantzen and Henri – and this is what they had signalled they would do. Both sides, then, would appeal. On our side, the appeal was only worth going ahead with if we believed there was some new evidence to bring, which Olivier felt there was.

Right at the end of the trial, he had introduced some eye-witness testimony that had asserted that Concorde had been on fire *before* it had encountered the piece of wear-strip off the Continental DC-10. The prefect and the prosecutor and the court had dismissed this, however, because they had countered that the accounts were both unreliable and contradictory. Particular attention had been placed by the prosecution on discrediting the testimony of two firemen who had watched Concorde take off from a vantage point at CDG.

The men had been standing next to each other. One had served at the airport for many years and was considered highly experienced; the other had not served at CDG as long and was therefore less so.

The more experienced of the two had testified he'd seen flames coming out of the left side of the aircraft long before AF4590 had hit the patch of runway where the wear-strip had been found; the other had said the flames had come out the right side. Because the fire had actually broken out under the port – left-hand – wing, the testimony of both men had been dismissed.

Olivier had made the case that it was not just a question of experience but of perspective.

The more experienced of the two had described the fire as being on the left because he had been describing the *actual wing* – the port wing. His less experienced colleague had been describing the position of the fire from his *point of view* – seen from the front, which is how they had viewed AF4590 taking off, the fire *had* been on the aircraft's right side. When this discrepancy was pointed out, along with an assurance by Olivier and his team that there was more testimony that was relevant – the case for a retrial was accepted.

The additional testimony bolstered a key assertion that AF4590 had been on fire before it had hit the wear-strip. Could this really be, I wanted to know.

Olivier and his team took me through the new data. One

of the things that hadn't received due attention in the trial, they pointed out, but which we had been well aware of in the investigation, was what appeared to have been an explosion, referred to as an 'overpressure', in the vicinity of the Number 2 Engine 'dry bay' shortly after take-off. Dry bays were compartments above each one of the four engines that housed the Olympus's electronic controls.

Dry bays do what they say they do. They're supposed to be free of fuel or other liquids. After the crash, inspection hatches for Dry Bay Number 2 had been found in a field just beyond CDG's perimeter fence. The implication was that the hatches had blown off – possibly due to fuel 'misting' in the bay, or its corresponding fuel tank, then exploding. If there was misted fuel in the dry bay, though, where had it come from? Misting didn't occur until things were almost empty. The dry bay should have been totally dry, while the adjacent fuel tank – Number 2 – would have been practically full at this stage of flight.

Something else that had been cursorily dismissed in the original trial now became more relevant.

At the crash site, most of the needles on the gauges of the fuel tanks were where you'd have expected them to have been given the aircraft's brief flight history. Despite the high-g impact, the fact that they seemed to have accurately reflected the fuel status of each tank at the moment of impact indicated to the investigators the readings ought to be considered accurate.

The two discrepancies were the gauges for Tank Number 5, which had shown plenty of fuel – 2 tonnes – and Tank Number 2 – the feeder tank for Engine Number 2 – that had shown hardly any: 0.1 tonne.

This, as Olivier had raised in the first trial, didn't make

sense. Tank Number 5, the tank that had been holed by the rupture after it had been struck by the rubber from the tyre, ought, by rights, to have exhibited little to no fuel.

Tank Number 2, which hadn't been struck, should have been almost full.

And yet, the opposite was found to be true.

But since Dry Bay Number 2 was above Tank Number 2 – and Dry Bay Number 2 had apparently suffered an explosion as the aircraft crossed the airfield boundary – what series of events could have led to it having been on fire in this way?

Tanks didn't just burst into flames spontaneously. The accident investigation had made no mention of holes from the burst tyre in Tank Number 2.

Did this, then, merit taking another look at what the eyewitnesses had reported – Concorde having been on fire *before* it had ever hit the wear-strip?

In conjunction with other evidence, Olivier and his team believed it did.

Amongst it were some anomalous readings on the flight data recorder as the aircraft – by now well into its take-off run – had crossed a 'ridge' that marked the division between the main section of runway and an extension to the threshold, where the aircraft started moving. CDG had lengthened the runway some years earlier by laying down additional tarmac at the take-off end.

The ridge at the point where the new and old sections met wasn't as smooth as it should have been – in fact, it was notoriously rough; like your car hitting a speed bump really fast.

Concorde had always used the full length of CDG's Runway 26R and, on 25th July 2000, the flight data recorder showed that it went over the bump at around 100 knots. At that moment, the FDR also recorded that there was a plateauing in longitudinal acceleration and a spike in lateral acceleration – in other words, the aircraft had kicked sideways. This had corresponded with a strange sound picked up on the cockpit voice recorder at that same moment – a *whoosh* sound. This was the noise I had noticed when listening to the CVR at the DGAC's offices in Paris. The *whoosh* had appeared on the recording fully 700 metres before the point that the wear-strip had been.

When I'd first heard this on the cockpit voice recorder – way back in the early days of the crash investigation – the noise had registered, but not as anything I knew. Now it did. Chris and I had recently taken Amy to a big firework display at a local theme park. It was the sound a rocket makes at the second of ignition, or the starburst of a large firework in the sky. The noise had baffled investigators, so it had been set to one side – as an anomaly. But it was clearly, Olivier said, a pyrotechnic sound of some sort – like the sound fuel makes when it catches fire.

Put these three things together – the aircraft hitting 26R's 'speed bump' at the exact same moment as the *whoosh* heard on the CVR, and with eyewitnesses saying that the aircraft had been on fire *before* it had encountered the wear-strip – and maybe we had something, Olivier said.

Maybe we did. This theory was compounded, we began to realise, when it emerged that there were many more witnesses than the two firemen who had said that the aircraft had been on fire before hitting the wear-strip.

Eyewitness reports by people who had seen Concorde take off had been collected by the *gendarmerie* shortly after the accident. These reports had been submitted to the prosecution but had been dismissed because, taken together, they didn't add up. *En masse*, the witnesses' accounts had placed

Concorde at different points on the runway when they had first seen it on fire, so the prosecution had adopted the same very literal position it had adopted with the firemen – that this inconsistency rendered *all* the testimony unreliable and should therefore be thrown out. Which, indeed, it had been.

Olivier and his team had submitted a different line of argument in their appeal.

The reason the accounts hadn't added up, they said, was not because they had conflicted with one another, but because there had been gaps in what the witnesses had perceived. Some had had their view partly obscured by buildings. Others had said other aircraft had been in the way for critical moments of Concorde's take-off roll.

All in all, the *gendarmerie* had taken about thirty statements by people who said they had seen AF4590 before it had hit the wear-strip.

After whittling them down, it had gathered as many as eighteen reports from witnesses who said that Concorde was already on fire by this point.

Olivier's team now had an inspirational idea: was there a way of tying the viewpoints together to see if they exhibited a kind of 'group perspective'?

Experts were brought in to assemble a virtual reconstruction of the aircraft's take-off run, using 3D, computer-generated modelling techniques.

The model used every available piece of data – from the flight data recorder, the cockpit voice recorder, GPS, physical evidence on the runway – soot and fuel marks, for example, showing where the tanks had ruptured and the fire had first broken out – communications from air traffic, ground radar tapes, and many others. *Let's tie all this data together,* they said, *and enter all those obstacles that had obscured people's views and overlay* that *with the positions of those eighteen eye-witnesses to see if it added up to any kind of pattern.*

It did.

The reason Observer A's viewpoint had differed from Observer B's was because they'd been unable to see Concorde on fire at the same time.

The eyewitnesses were bound, too, by some other factors: they were professionals – almost all had worked in and around the airport – and they had made a conscious *choice* to watch AF4590 take off. Most of them had watched Concorde take off many times before and knew what to expect. These weren't normal, error-prone eyewitnesses – they were *observers*.

And what they had seen coming out of the back this time was different – not the flames of reheat, but the flames of an aircraft on fire. An external fire under the port wing before the aircraft had encountered that piece of metal.

If the aircraft was on fire before it hit the wear-strip, what explained it?

We had debris from the Number 2 Dry Bay somewhere beyond the end of the runway.

We had a fuel gauge for Tank Number 2 that had registered almost empty.

Instead of misting in the Number 2 Dry Bay, it wasn't inconceivable, therefore, that there had been an explosion in *Tank* Number 2, as it had emptied – and that the overpressure had blown the hatch on the Number 2 Dry Bay above it.

What might have caused the Number 2 Tank to rupture and explode?

Suspicion fell back on that anomalous reading from the flight data recorder as the aircraft had hit the ridge marking the runway extension – the lateral jump it had made to the left coincident with the *whoosh* on the CVR.

And at that point, our focus veered back on to an old suspect from the investigation – one that had again been dismissed in the original trial: the spacer that had been missing from the left-hand undercarriage assembly.

As a working theory, this is what we started to run with. We knew that with each take-off and landing that had taken place with the spacer missing – those two flights to New York and back – the shear bushes for the wheels had migrated further and further apart, because we had the witness marks on the undercarriage to prove it. The left-hand bogie assembly, therefore, was already in a dangerously parlous state when AF4590 began its take-off run. When the wheels hit the ridge at 100 knots, there was nowhere left for the bushes to go, so they broke, a piece of undercarriage flew off and hit Tank Number 2.

Tank Number 5 – the tank that had been ruptured by the stresses imposed on it when the rubber hit the wing – was forward of the left undercarriage assembly; Tank Number 2 behind it. It made more sense that a piece of debris flying off the wheel would strike a piece of wing behind, rather than in front of it, although there would have been nothing – as evidenced by the rupture – to have stopped a piece of rubber shocking the wing beneath Tank Number 5 either.

So, with fuel now streaming out of the Number 2 Tank, what would have caused it to ignite?

In the original scenario – fuel streaming back from Tank Number 5, just aft of the wing leading edge – the UK investigators had rejected the idea that the reheats had ignited the fuel because of the time it would have taken for the fire to have travelled from the afterburning section of the engine up the fuel stream. They'd worked this out from fuel deposits on the runway and soot marks some way further on where the fuel had ignited. The distance between the two wasn't far, but enough for them to work out how long it would have taken for the fuel to catch fire – and the maths simply didn't add up for Tank Number 5, which is why the Brits had adhered to the view that the ignition must have been caused by sparks from broken wiring in the undercarriage bay.

But this new theory changed things. The distance between Tank Number 2 and the reheats was much shorter because the Number 2 Tank was much closer to the wing trailing-edge and the engine exhaust nozzles. We were talking fractions of a second's difference here, but fractions of seconds were what the boffins dealt in. What made it significant was that it tied in with what the eyewitnesses said they had seen – the aircraft on fire before it had hit the wear-strip.

It also provided a potential solution to another question that had vexed the investigators: how the wear-strip, which had fallen off the DC-10's right engine, had somehow migrated to the other side of the runway to strike a wheel on AF4590's left undercarriage. The accident report had ducked the issue – a fudge made more incomprehensible by the fact the wear-strip had migrated back to the right-hand side of the runway by the time investigators found it.

It had put me in mind of the 'magic bullet' that had supposedly killed JFK.*

The only thing that sat a little uncomfortably with us was

---

* The so-called magic bullet theory said that the single bullet that had supposedly killed President John F. Kennedy struck him in the neck and exited through his throat, whereupon it entered the chest of Texas Governor John Connally, who was riding with Kennedy, went through Connally's wrist and embedded itself in his thigh.

that we were now talking about two separate failures. But this was consistent with the data.

The wing skin covering Tank Number 5 had been hit by a large piece of rubber from the tyre that had exploded when it had hit the wear-strip, causing the tank to rupture from within. Tank Number 2, however, had been punctured *beforehand* by a piece of undercarriage debris thrown up when the left-hand wheel assembly – weakened by previous take-offs and landings because of the missing spacer – struck the ridge between the two sections of runway.

This was consistent, too, with the fact that the alarm from the flat-tyre detection system that had been installed some years previously hadn't activated.

What *had* happened, however, was that the co-pilot Marcot, shortly after calling V_I – the speed at which they had been committed to take-off – had been heard on the cockpit voice recorder to say, *'Attention!'* – 'Watch out!' – to Captain Marty for a reason no one had ever been able to explain. It had coincided with the jump to the left logged by the flight data recorder when AF4590 hit the ridge and, afterwards, with Marty's inexplicable decision to rotate 15 knots early.

It wasn't our intention to raise the matter of the early rotation because the data appeared to show that AF4590 had been on the runway centreline when he had initiated it. But, all the same, it was hard not to speculate – amongst ourselves, at least – about an alternate scenario: that hitting the ridge had caused the trolley-wheel to stop wobbling and lock hard left; that AF4590 had started to veer off the runway; and Marty had had no choice but to rotate 15 knots early because of the unthinkable alternative: hitting the just-landed 747 of the President of France.

Be that as it may, this new argument was what Olivier and

his team would present to the court. And it was right that they would – because it was supported by the evidence. Unfolded in the right way, we had no doubt that it had the potential to carry the day: to exonerate Continental, and John Taylor.

But it had a corollary – one that saddened me deeply. The evidence supported what we'd always said – what every air crash investigator knew: that almost all aircraft accidents happened through a multi-causal chain of events, not because of a single point failure. But the flipside to this argument was that my aeroplane must have had some inherent vulnerability all along.

47

The appeal opened in a courthouse in Versailles on 8th March 2012. The court decided that two of the defendants – Henri Perrier and Jacques Hérubel – would be tried separately on account of Henri's failing health; years of worry and stress had taken their toll and Henri, eighty-three years old, was in and out of hospital for treatment. The court ordered that it could not wait for him and that he and Hérubel would be tried later. A date was set for the following year.

Two months into the appeal, I made my way across the Channel to present my testimony.

I found myself in a room not unlike the one I'd been in before. The seats of the defending and prosecuting lawyers were again arranged like the stalls of a choir, facing each other in front of the altar-like bench where the judges sat – a mixed panel this time, presided over by a female president.

Over a period of several hours, I was examined and cross-examined on the evidence outlining why AF4590 had been on fire before it had hit the wear-strip. Once again, it was Olivier Metzner who acted as my inquisitor.

'Why did we not hear this from you in the first trial?' he began.

'Because the evidence wasn't available,' I replied, speaking to the judge. 'This analysis has only become possible thanks to the extraordinary work done in fusing the eyewitness testimony into a single picture via a very comprehensive digital simulation.'

'And what did it show you?'

'A considerable consistency between eighteen witnesses, who said that the aircraft had been on fire before it reached the wear-strip.'

'How certain are you of this, *Capitaine* Bannister?'

I looked at the judge. 'I'm convinced, Madame La Présidente. All the witnesses say they saw the aircraft on fire after it hit the ridge of the runway extension. What the simulation proves, to my satisfaction, is how these accounts align. AF4590 was on fire 700 metres before it hit the wear-strip.'

Olivier arched an eyebrow as he looked at the judge and asked me to sum up the various pieces of the causal chain that had precipitated this event.

I took a deep breath.

There was the crucial omission of the spacer.

The fact that the left wheel assembly had been shocked into partial disintegration when it had hit the ridge.

The fact that a piece of debris had shot into Tank Number 2 behind the undercarriage, causing fuel to stream through its punctured skin – and the fuel's ignition by the reheats of Engines 1 and 2.

And the fact that a piece of rubber from the right-hand tyre of the front set of wheels on the left bogie had struck the forward part of the wing under Tank 5, sending a shock wave through the fuel that had ruptured it from within.

When I finished, I sat down beside Olivier and his team.

Everyone agreed the way we had presented the new evidence – helped by our depiction of the events as they had unfolded from the point of view of the eyewitnesses via our 3D digital reconstruction – had gone well.

But over the next few days, the prosecution tried to

undermine it by saying that eyewitness testimony was unreliable under such circumstances because startled people didn't mentally record events in the way they really happened – the shock of what they were witnessing twisted and contorted them into something unreal. I kicked myself when I realised this was a plank in their argument because I had meant to address it during my testimony.

I had rehearsed the argument carefully. These were not the kinds of random, startled witnesses you got to a car accident or a terrorist outrage.

All of our eyewitnesses were aviation professionals who had *chosen* to watch AF4590 take off. They were observers. Concorde was that kind of an aeroplane. No matter how many times you'd seen her before, you stopped what you were doing and watched – you *observed*. This is what *these* people had done. Almost all of them had seen Concorde take off before, so what they saw wasn't new. What they *witnessed*, however, was.

On 29th November, almost eight months after the appeal began, the court delivered its verdict – it overturned the result of the previous trial, absolving Continental of all criminal responsibility. It also cleared John Taylor.

It did, however, uphold a ruling that Continental bore responsibility for the disaster and therefore should pay Air France the one million euros it had been ordered to pay previously for damage done to Air France's reputation.

It did not challenge the first court's ruling that tied the wear-strip from the DC-10 to the chain of events that caused the crash. That there had been a 'chain of events', however, as we'd stated all along, was no longer disputed.

I watched on TV as Olivier stood outside the court, clearly enjoying his moment in the wintry Paris sunshine. The ruling by the court had put to an end twelve years of 'wrongful accusations', he told a bank of waiting reporters.

The decision was welcomed by professional aviation bodies around the world, who saw it as a restoration of sanity – a reversal of a creeping trend, not just in France, to pursue criminal charges against pilots and other aviation professionals under such circumstances, instead of pursuing the truth. Only by *getting to the truth*, bodies like the International Civil Aviation Organisation said, would aviation eliminate the kinds of errors that had led to this terrible event – an accident, we all needed to remember – that had killed 113 people.

For me, the result was a justification for all the work that had gone into both the trial and the appeal over the previous five or so years of my life.

It was, too, a partial exoneration of Concorde.

The appeal court had ruled that Concorde had been vulnerable to what it called 'shock'; that officials had known about these vulnerabilities and it should not have been allowed to fly with them. It also ruled that although they had missed chances to rectify these flaws, they could be 'accused of no serious misconduct'. Or, as the US Flight Safety Foundation had managed to sum it up, that professional human error didn't amount to criminal conduct.

A certain bitter taste persisted, however. Absent from the appeal court's findings was any meaningful discussion of the many institutional errors at Air France – exposed by the UK side of the crash investigation – that had led, directly or indirectly, to the crash. Right to the end, the BEA insisted that its report had reached the right conclusions, even though so much of its evidence had been challenged – and upheld – in the appeals court. The charge by one of its officials that during eighteen months of its investigation I had never raised my theory of the 'unauthorised procedure' that precipitated the crash – overfilling the tanks (although there were others) – was simply untrue. On the UK side, we brought it up time and again at our meetings with them.

If it had been possible for me to add to the court's conclusions the kind of rider that the British accident investigation authority had added to the BEA report, then I'd have done so.

No such opportunity existed, of course. *Madame La France* had spoken.

But we had won; that was all that really mattered. The people we'd gone into court to defend against the lunatic charges brought against them had emerged free.

The 'shock' charge against Concorde hadn't surprised me as much as it seemed to have surprised some of those who wrote about the verdict of the appeals court. I, too, had had to come to terms with the idea of 'vulnerability' – what it meant in this context. Those of us who knew Concorde had always accepted she was a thoroughbred. Thoroughbreds aren't like other breeds. They have their idiosyncrasies. To fly where she did in the way that she did, Concorde had had to have a special wing. Its skin had had to be thin. But that hadn't made her dangerous. Over the years, as problems had emerged – problems like the tyres – we'd fixed them; at least, at BA we had. Concorde wasn't like other aircraft. She was special and she'd needed to be respected in that way. Like Bill Weaver's SR-71, our aeroplane had operated on the 'edge of possible'. *Respect the airplane and she'll respect you back*, he'd said.

It made me thankful that, in the UK, we had celebrated Concorde in the way we had, when we had. An image flashed through my mind of the moment I had looked up through her visor as we'd flown down the Mall and I'd first glimpsed the people and the bunting and Buckingham Palace below. I felt, briefly, the physical sensation of lift the aircraft had received from the crowd.

Concorde had been correctly remembered – not just for her part in aviation history, but in the hearts of people for whom she had been an icon.

My overriding regret was that Henri Perrier hadn't lived to know any of this. Sadly, he had passed away shortly before the verdict of the appeal court had been delivered.

Three months later, we lost another. On 17th March 2013, Olivier Metzner was found drowned in waters off the island he owned in Brittany.

Because of his defence of dictators and high-profile criminals, Olivier had been known as the 'gangsters' lawyer'. Police found a suicide note. Many, though, found it hard to shake the belief foul play had been involved.

Somewhere along the line, people said, he had taken on a case too far.

48

Brooklands Museum, just inside the M25 London ring road, in leafy, suburban Weybridge – birthplace of British motorsport and British aviation – on a sunny summer's morning ten years on from the events in that Versailles courtroom.

Strolling from my car to the Concorde simulator, I glanced, as I always did, at the gathering of Vickers-built aircraft in the outside exhibit area. They included a Viscount, one of the types that had caught my eye all those years ago high above Bournemouth Beach, a VC10 and Concorde *Delta Golf*, the aeroplane we'd shipped here a decade-and-a-half earlier from Bristol.

Delta Golf was – and remains – the centrepiece of the museum's 'Concorde Experience'. After arriving in 2004, she opened to the public following extensive restoration in 2006. The Concorde simulator, after its extensive restoration, was opened to the public here in 2009.

My association with Brooklands had begun in 1991, when, as head of communications for flight operations at BA, I'd got a note from my boss to call the then-chairman, Lord King. My first thought was that I must be in some kind of trouble, even though my boss assured me I wasn't. Shortly after I'd responded to the message, Lord King rang: 'Mike, I've got a favour to ask.'

'A favour, Chairman?'

'There's a VC10 fuselage at the Brooklands Museum that's

in a bit of a bad state of repair. Can you organise for some-one to get it repainted for me?'

'Yes,' I said, 'of course.'

It was only afterwards, I thought about what I'd just prom-ised. Why me? Why hadn't the Chairman called Engineering? What did I know about repainting a VC10?

The fuselage in question had been a forward section that had been used on-site in pressurisation tests way back when – before I'd ever set foot in one. It had been kept outside and had deteriorated badly. I called in a few favours and managed to gather a posse who got it repainted in its original BOAC colours. By the time they had finished, it looked absolutely fantastic.

It was thus that my association with Brooklands began. I became a corporate member – BA held some corporate events there – then *Delta Golf* arrived.

The aircraft's restoration at Brooklands coincided with the years that immediately followed my retirement.

Not long afterwards, Lord Trefgarne, the museum's chair-man, asked if I might consider becoming a trustee. I jumped at it.

I had been looking for something aircraft-oriented to do after BA – and a trusteeship at Brooklands, with its long Concorde association – not to mention its Mecca status for automotive and aviation innovation at the dawn of these two industries in the UK – suited me perfectly. As someone who'd been lucky to live his dreams in a thirty-seven-year career, a direct association with Concorde at Brooklands felt wonderfully full-circle.

On this summer's morning, I was making my way to the simulator room to fly with a group of children who'd come to the museum on a special school trip.

Somebody – knowing I was here for a meeting of the trustees – had asked if I'd reprise my one-time role as a Concorde training captain for them.

Because the simulator was the real thing – this was the simulator on which I and every other Concorde pilot at BA had trained to fly the aircraft – it flew identically to the real thing. And since it had been put back together, with a visuals system as good as anything that virtual reality had to offer, flying the sim was breathtaking. I had acted as 'instructor' on lots of flights for guests and visitors and the look on their faces afterwards was always rewarding.

We'd take off, fly around a bit, then do some stupid stuff – flying low-level past well-known landmarks and under bridges – before landing back at an airport: Heathrow, New York, Hong Kong or Sydney. We'd had a lot of people fly it, ranging from an eighty-five-year-old lady who'd never flown in her life to a Boeing test pilot who'd made the trip to Weybridge all the way from Seattle.

But there was something about seeing a child's face after they'd had a go that was pure magic.

Part of Brooklands' charter is to inspire young people to consider adopting STEM subjects – Science, Technology, Engineering and Maths – subjects that would enable them, if they so desired, to take up a career in the automotive or aerospace industries.

Since it had opened 'The Concorde Experience', a very realistic 'virtual flight' aboard *Delta Golf* had been a big part of this.

By 2017, in just eleven years, over 40 per cent more people had 'flown' on the Brooklands Concorde than had ever flown on any BA Concorde during her twenty-seven years of service. A great many of those hadn't even been born when Concorde had retired.

I remembered what it had been like for me, as that seven-year-old, gazing up from that Bournemouth Beach at those silver specks flying east.

I thought about my first solo flight at Luton, then heading off, after school, to Hamble. The lucky breaks that had got me into the right-hand seat of a VC10 early on in my career – and my tuition at the hands of a cadre of people who would soon all be gone: pilots and navigators who, by some miracle, had survived bombing raids over Nazi Germany, night after night.

Men like Geoff Morrell, Chopper Knights and Arthur Winstanley who had put me through my paces over the Atlantic west of the Cliffs of Moher.

And then, the Concorde years. It doesn't matter what a child's dream is – mine happened to have been aviation. The thing is to give every child the opportunity to achieve theirs. And the key to that, of course, is education.

The love for Concorde, that sense of ownership that people felt, remained strong. And not just amongst Brits, but people of all nationalities.

For those who were familiar with the supersonic transport story, I saw something else in their eyes, as well: pride – the same pride people of a certain age have for the Apollo Moon programme. A pride that says, 'Look what humanity can do when we get behind something.' It was a pride I felt.

When I entered the simulator room, the children were there, waiting.

A girl who wasn't much older than our Amy had been the day that Concorde retired – she had been nine or ten back then – stood up and shook my hand awkwardly. I found out from her teacher she was first in line to fly.

'Hey,' I said, 'do you want to go and fly to the edge of space?'

Her eyes lit up and her shyness evaporated. She looked at her teacher, then back at me. 'Me?'

'Yes,' I smiled, 'you.'

It took around an hour for them all to have a go – and along the way they'd fired me the usual questions:

Had Concorde really flown faster than a bullet?

Why was this amazing plane now in a museum?

Has a new Concorde been invented?

Did you have to wear a spacesuit to fly in it?

I answered them as best I could. Some had been easier than others. Some had made me smile. Others had filled me with a longing for the past.

After they'd left, on my own again, I sat back down in the right-hand seat. She was still on the runway at Heathrow, where the last child, with a bit of help from me, had landed her and come to a full stop. In the silence around me, I fancied on the ether that I heard a once-familiar voice:

'OK, Mike, take off, straight ahead and level off at 2,500 feet – if you can.' *Keith Myers, my old training captain . . .*

If I can? I'm an experienced pilot with twenty-two years on Concorde. What do you mean, if I can?

Throttles to the stops, reheats in my back, brakes off, that phenomenal acceleration, rapid, mental overload and, *boom*, suddenly, I was at 4,000 feet.

Only, this time, I didn't stop. I kept on climbing.

Up into that band of thin air, up, up, to where the blue met the inky blackness of space.

Epilogue

Concorde remains the only successful supersonic airliner ever built.

She clocked up more supersonic flight hours than almost all the world's air forces.

She enabled you to travel with ease and in style from London to New York and back in a day – with more than four hours in New York for business.

And, along the way, she allowed me to log a small Guinness Book of Records-type achievement as well: whilst we didn't count 'trips' but total hours, I managed to clock up more Concorde hours than anyone else – just under 10,000, of which 7,000 were supersonic, which equates to thirty-four round trips to the Moon.

Do I miss her?

Of course. Desperately, really. But I've had plenty to distract me in the years since we both retired.

In addition to my vice chairmanship at Brooklands, I am chair of governors at a local independent school, have a couple of corporate non-executive directorships, am active in our local church, am chair of a London property company, former executive chair of a food bank, secretary of an airline cabin crew support organisation, and president of our local Air Cadet squadron – a wonderful organisation that has been dear to my heart ever since Amy learnt to fly with them.

I run my own aviation consultancy specialising in airline management, test flying, operational safety and security matters across a range of aircraft types and geographical locations.

Plus, I do my very best to raise much needed funds for deserving charities.

In short, I'm enjoying life – the life that I envisaged when I retired. Chris and I see friends and we travel; and I even occasionally get in a round of golf.

Regrets? Yes, I have those too. Early in my career, I was conscious my job took me away from my family far more than I'd have wished – that I hadn't spent as much time with my kids as I'd liked to have – still a source of regret.

My two children James and Robbie, from my first marriage to Maggie, are wonderful sons. Unsurprisingly, perhaps, neither of them ever exhibited any desire to go into aviation. If you'd asked me when they were born if I'd liked one or both of them to have become pilots, I might have said yes. But with the benefit of hindsight, as a parent, none of that has mattered. I care, of course, that they are happy, but the bonus for me is that they are both very good at, and happy in, what they do. One of my most special memories is of the time I flew them in *Alpha Foxtrot* from Heathrow to Paris, where Concorde had a starring role in the 1997 Le Bourget air show. We spent a great day in the sunshine, watching the flying, sitting out on one of *Alpha Foxtrot*'s wings – something that the health and safety gurus prohibited very shortly afterwards.

As for the future, I don't believe *Homo sapiens* takes backwards steps for long.

Or that the Concorde story is over yet.

The question I still get asked more than any other is *why*: why was a thing of such beauty, such an innovative leap, put out to pasture when she was?

I have always remained convinced she'll have a successor – one of the reasons I have kept engaged with various proposals that are out there for supersonic business jets and airliners.

The one very likely to succeed, it seems to me, is a design called 'Overture', developed by the American start-up, Boom.

When I first came across the company on a visit to the Farnborough Air Show several years ago, I was shown a concept model with a planform that included a highly swept wing and a sharp, almost fighter-like nose. It was very purposeful looking.

Two years later, I returned to the show and saw an iteration that was quite different – a gorgeous-looking design with wings that had an S-shaped leading-edge, plus long, angular under-wing intakes and a familiar vertical fin.

When I pointed out its strong resemblance to Concorde to its chief designer, he smiled and said: 'Yeah, well, you Brits got it right first time.'

We should, of course, in the interests of accuracy and completion, include the French here too.

And for posterity this as well: for all the rancour that existed between the UK and *Madame La France* during the Concorde crash investigation and the trial, I'm delighted to say that that has all gone. In its place, everyone I know chooses to remember the remarkable cooperation between our two countries at a time when international collaborative aerospace efforts didn't exist – Concorde being a world first in that department as well. Without that collaboration, not only would she never have been built, but Europe would be without the phenomenally successful aerospace industry it enjoys today; an industry that brings together tens of thousands of employees across the Continent, as well as in the UK; an organisation that's responsible for some of the best engineering on the planet, a family of aircraft that is successful, safe and reliable, and which is more than a match for Boeing. I'm talking, of course, about Airbus.

And when I think of Airbus, I choose also to remember my dear friend, Henri.

A Boom demonstrator aircraft, the XB-1 Baby Boom, may well have flown by the time these words appear in print. The Overture is slated to be in service by the end of this decade.* Carbon net-zero with sustainable fuel, a range of 4,250 miles, a Mach 1.7 cruise speed, up to eighty passengers and a cruising altitude of 60,000 feet.

As for Amy, still only in her twenties, she is currently serving as a senior first officer and instructor on the Boeing 737 for a major UK carrier.

She always said she wanted to fly a supersonic successor to Concorde and in Boom's Overture, or something like it, I truly believe she will get her wish. I most certainly hope so. Maybe, just maybe, they'll let me have a go too.

---

* At the Farnborough International Air Show in July 2022, Boom Supersonic announced a design change, giving the Overture four engines instead of three, a gull wing and a contoured fuselage. This final iteration of the aircraft is expected to go into production in 2024. See https://boomsupersonic.com/overture

Acknowledgements

On the 16th of November 1985, we took Concorde G-BOAG on a special flight to Aruba, one of our destination islands in the southern Caribbean. It was a trip to celebrate the opening of the 'Concorde Hotel and Casino' there. As dusk approached, I can clearly remember standing on the airfield looking out at this beautiful aircraft bathed in glorious Caribbean sunshine, with the tropical thunderstorms that were typical of the place rolling around in the distance.

In a few moments, it started to rain heavily, and Concorde was framed in the most glorious of rainbows. With the sun shining on my back there was an idyllic image as the rainbow stretched from the aircraft's nose to its tail.

It was then that I realised that something as beautiful as a rainbow is not made up of the simple seven colours that we all learn about as children. It is made up of thousands, if not millions, of shades. It is not complete without each and every one of them. Likewise, this book is not just my work. I could not have done it without the assistance and support of so many people. In listing my acknowledgements and thanks, I know that I will miss out some, and for that I apologise. But here goes . . .

To my parents, Joan and Arthur, who had the courage to believe in the seven-year-old child's dream, and to help it come true. My brother Keith, and his wife, Jan, who tolerated this younger sibling's obsession, and helped in every way that they could.

To Nick Cook, Rowland White, Mark Lucas, Nick Lowndes,

Kay Halsey, Ruth Atkins, Eloise Austin, Jen Harlow and Adrian Meredith. I had no idea how many dedicated people would be involved in bringing my vague idea of writing this book to fruition. During its writing, I was also very grateful for a number of online data sources, most notably www.concordesst.com and www.heritageconcorde.com, which were especially good on the history of the aircraft and on the minutiae of certain technical details.

To all of those who spent the time to share their aviation and Concorde knowledge with me over my fifty-six years as a pilot. If, as an instructor, I've passed on just a small percentage of their wisdom to others, then I've done my job.

To the people who made Concorde the exceptional aircraft that it is. The 'Mach 2 Gee-Whizz' stuff is great, but it's the people who really made the legend. All of those in the British Airways 'Concorde Family', so many of whom have become close friends. The two-and-a-half-million people who flew on the BA Concorde during its twenty-seven years of operations. The public who paid for the research and development and had the love and admiration for the aircraft to stick with it through thick and thin. The folks at Brooklands, both when it was the place where 33 per cent of each Concorde was built, and now as the museum that hosts the largest Concorde collection in the world.

To those in my life who are no longer with us but are essential to this tale, including Maggie Bannister, Henri Perrier, Brian Trubshaw, Geoff Morell, Bob Knights and Tony Meadows. Plus, of course, the memory of the 113 souls who perished at Gonesse.

And, perhaps with the most important last, my family.

Chris, James, Robbie, Amy, Sue, Rhiannon and William. They have been so patient in understanding my passion for this aircraft, and for tolerating the times that it has taken me away from them.

I love them all dearly and literally could not have done this without them.

Picture permissions

p.1, top, Wikimedia Commons; middle, Airbus; bottom, BAe

p.2, top, Shawshots / Alamy Stock Photo; upper middle, Pan Am; lower middle, Victor Drees/Daily Express/Hulton Archive/Getty Images; bottom, Arthur Gibson

p.3, top, Source Unknown; middle, Fred Willemsen; bottom, Airbus

p.4, top left & right, middle, Author's own; bottom, BA

p.5, top, BA; middle, Author's own; bottom four, Charles Skilton

p.6, top, Greater Orlando Aviation Authority; middle left, Adrian Meredith; middle right, PA/Alamy Stock Photo; bottom, Braniff Airways Foundation

p.7, top, Author's own; middle, Adrian Meredith; bottom, Iain A Mackenzie

p.8, top, c/o Mike Bannister; middle right & botttom, Author's own; middle left, Source Unknown

p.9, top, Adrian Meredith; middle, Crown Copyright; bottom, Author's own

p.10, top left & central image, John Dibbs; middle left, c/o Mike Bannister; bottom left, Source unknown

p.11, bottom left, Adrian Meredith; bottom right, Arthur Gibson

p.12, top, Toshihiko Sato/AP; middle left, Source Unknown; middle right, Bae; bottom right, Author's own; bottom left, Air France

p.13, top, PA Images / Alamy Stock Photo; middle, Source Unknown; bottom right, Adrian Meredith; bottom left, Author's own

p.14, all, Author's own

p.15, top, Anthony Kay @ airliners.net; middle, Trinity Mirror / Mirrorpix / Alamy Stock Photo; bottom, BA

p.16, top, aviation-images.com/Universal Images Group via Getty Images; middle, Gerald Ramshaw

Endpapers: Adrian Meredith
Cutaway image of Concorde: Mike Badrocke

Every effort has been made to trace copyright holders and to obtain their permission for the use of copyright material. The publisher apologizes for any errors or omissions and would be grateful to be notified of any corrections that should be incorporated in future editions of this book.

Deeper Insights

AAIB Air Accident Investigation Branch – the UK aircraft accident investigation authority.

Ab initio A pilot, at the beginning of their training process, who has very little flying experience.

Accel point The specific geographical point at which Concorde could start its acceleration through Mach 1, the speed of sound.

Accelerometer An instrument for measuring the acceleration of a moving or vibrating body.

Actuator A device that, on receipt of a signal or input, produces motion using energy.

Afterburner *See* Reheat.

Aileron A movable part of an aircraft's wing that is used by the pilot, or autopilot, to rotate ('roll') the aircraft around its longitudinal axis – one wing going up, and the other down.

Air intake The most critical part of the powerplant and, perhaps, Concorde's most important technical feature, the air intake was an approximately eleven-foot-long assembly in front of each Concorde engine. It managed the incoming air so that the Olympus 593 could handle it during all phases of operation, including take-off, subsonic and supersonic flight.

Air intake ramp (door) One of two computer-controlled, hydraulically powered, variable position ramps or doors, which were hinged on the upper surface of the intake and modulated up and down to adjust the intake inlet area and create carefully managed shock waves within the intake.

Alpha Golf, **etc.** The last two letters of a British Airways Concorde registration in the international phonetic alphabet. The seven operational Concordes were registered G-BOAA to G-BOAG. A is 'Alpha', B – 'Beta', C – 'Charlie', D – 'Delta', E – 'Echo', F – 'Foxtrot' and G – 'Golf'.

ATC Air Traffic Control – the organisation that co-ordinates and directs aircraft movements.

Auto-stabilisation A system to enhance the stability of an aircraft independently of pilot inputs. It uses sensors and computers to control the movement of flight control surfaces.

Axis/Axes (of an aircraft) The axis that extends lengthwise (nose through tail) is called the longitudinal axis, and the rotation about this axis is called 'roll'. The axis that extends crosswise (wingtip through wingtip) is called the lateral axis, and rotation about this axis is called 'pitch'. The axis that extends up and down through an aircraft (bottom to top) is called the vertical axis, and rotation about this axis is called 'yaw'. *Also see* Pitch, Roll and Yaw.

Back of the drag curve A drag curve is a graphical representation of drag plotted against speed. For a given weight, the minimum drag speed is the lowest point on that graph. Once below that speed, drag increases as speed reduces. Hence more power is needed to maintain a given speed. This characteristic is known as an aircraft being 'on the back of the drag curve'. On approach, Concorde was 'on the back of the drag curve' below about 265 knots.

BAe An abbreviation for British Aerospace, formed in 1977 by the merger of the British Aircraft Corporation, Hawker Siddeley Group and Scottish Aviation. It subsequently became BAE Systems in 1999 when it acquired Marconi Electronic Systems. As such, it is often abbreviated simply to BAE.

BEA – airline company An abbreviation for the British airline British European Airways, a forerunner of British Airways.

BEA – air accident investigation authority *Bureau d'Enquêtes et d'Analyses* – the French aircraft accident investigation authority.

Beech Baron A light, low-winged, twin-engine, unpressurised six-seat, US piston-powered aircraft. The 'D55' version was used for advanced flying training at the College of Air Training, Hamble.

Blackbird – Lockheed SR-71 Developed and manufactured by the American aerospace company Lockheed Corporation, the SR-71 was an initially secret, high-altitude, long-range, strategic reconnaissance aircraft. Capable of speeds well in excess of Mach 3, it was operated by the US Air Force, NASA and, in its 'Oxcart' variant (as the A-12), by the CIA.

'Black box' flight data recorder See FDR.

BOAC An abbreviation for the British airline British Overseas Airways Corporation, a forerunner of British Airways.

Boom Supersonic The trading name of Boom Technology Inc, Boom Supersonic is based at Centennial Airport, Dove Valley, near Denver, Colorado, USA. It is developing a new generation of supersonic passenger aircraft, called the Overture, designed to carry 65 to 80 passengers at 60,000 feet over distances of up to nearly 5,000 statute miles and at speeds up to Mach 1.7. It is also developing the Boom XB-1 'proof of concept' aircraft which may have conducted its first flight by the time that you read this.

Bowser A customised fuel tanker used to refuel aircraft.

C of L or C_L Centre of Lift – effectively the point through which all the lift on an aircraft is considered to act. It moves aft as speed increase and forward as speed reduces.

CAA Civil Aviation Authority – the UK national civil aviation regulatory and safety organisation.

'Capacity' For a flight deck crew member, 'capacity' is the ability to operate the aircraft whilst also thinking and planning ahead – to draw up contingency plans and leave sufficient mental and physical capability to handle emergency and/or unexpected situations, without becoming either overloaded or fully occupied.

Captain The pilot designated as being in command of the flight. The

Captain and the Co-Pilot are equally qualified and will usually alternate sectors as Handling Pilot and Non-Handling Pilot.

Certificate of Airworthiness Issued by the appropriate nation's regulatory authority, it is an aircraft's permission to operate in specific roles. It attests that the aircraft is 'airworthy' and conforms to its 'type design'. When the aircraft is registered, the certificate is issued in the name of the owner and in one or more different operating categories.

C of G (or CG) Centre of Gravity – effectively the point through which all the weight of an aircraft is considered to act. It does not vary with aircraft speed.

Chicago Convention Also known as the Convention on International Civil Aviation, it was established by the International Civil Aviation Organization (ICAO), a specialised agency of the United Nations which is charged with coordinating and regulating international air travel.

Chock-to-chock time Also known as 'block time', it is the period from the moment that the brakes are released at the beginning of a flight, to the time that they are reapplied after the flight.

Chuck Yeager Chuck (Charles Elwood) Yeager was the first pilot in history confirmed to have exceeded the speed of sound in level flight. This was achieved on 14th October 1947 flying the Bell X-1, *Glamorous Glennis*. He achieved Mach 1.05 at an altitude of 45,000 feet over Rogers Dry Lake in the Mojave Desert in California, USA.

'Clam-shell' doors (buckets) Part of Concorde's Secondary Nozzle system which was attached to the rear of each engine and a) modulated to optimise engine performance in subsonic flight, b) could be closed on landing to provide reverse thrust and c) could be deployed in flight (inners engines only) to significantly increase the aircraft's drag in order to rapidly reduce height or speed.

Concorde Treaty The Concorde Treaty, or Anglo-French Concorde Agreement, was signed in London on 29th November 1962. It built on inter-company agreements between Sud Aviation and the British

Aircraft Corporation (BAC) that were set up earlier in the year. The Agreement established development, risk and cost-sharing, the engine to be used and the desire to produce both medium and long-range versions of the aircraft. This was later abandoned as an objective. Key to the 'Treaty' nature of the Agreement, neither side could cancel without the approval of the other and, if one side pulled out unilaterally, it would remain liable for its share of the costs.

Conway engine With development starting in the 1940s, the Rolls-Royce RB.80 Conway was the first turbofan engine to enter service. It was used to power the Vickers VC10 and also the Handley Page Victor, Boeing 707-420 and the Douglas DC-8-40.

Co-Pilot The pilot designated as being second in command of the flight. The Co-Pilot and the Captain are equally qualified and will usually alternate sectors as Handling Pilot and Non-Handling Pilot.

CPL A Commercial Pilot Licence is the minimum qualification for a pilot to be permitted to operate a large passenger aircraft, and to be remunerated for doing so.

CRM Crew Resource Management is the structured procedure taught to, and used by, flight and cabin crews for improving aviation safety with a focus on interpersonal communication, leadership, and decision making in aircraft cockpits. Developed initially by David Beaty, a former BOAC pilot, it came under focus as a valuable tool following the 18th June 1972 *Papa India* accident at Staines, UK.

CVR A Cockpit Voice Recorder records and stores the audio signals of the microphones and earphones of the pilots' headsets and of an area microphone installed in the cockpit. As well as recording conversations it is used to identify actions both inside and outside of the cockpit, e.g. switch activations and external impacts. CVRs became mandatory in the UK following the 18th June 1972 *Papa India* accident at Staines, UK.

Decel point The variable point at which Concorde should start its deceleration to be below Mach 1, the speed of sound, by a specific geographical point.

Deep-stall A stall, usually of a 'T-tail' designed aircraft, where the effectiveness of the elevators on the top of the 'T' become reduced, or completely ineffective, because they are blanked by the stalled, turbulent air coming from the wing. Consequently, the elevators cannot produce enough nose-down pitching moment to recover the aircraft from the stall.

de Havilland Canada Chipmunk A tandem, two-seat, single-engine primary trainer aircraft designed and developed by Canadian aircraft manufacturer de Havilland Canada. It was developed shortly after the Second World War and sold in large numbers during the immediate post-war years, being typically employed as a replacement for the de Havilland Tiger Moth biplane.

Depressurisation drills Procedures and actions to be carried out by the flight crew should the aircraft cabin pressurisation system fail and lead to an effective cabin altitude in excess of 10,000 feet.

DGAC *Direction Générale de l'Aviation Civile* – the French national civil aviation regulatory and safety organisation.

Divergent Dutch roll On the VC10, without the protection of yaw dampers, high altitude Dutch roll is likely to occur. Above 25,000 feet, without pilot intervention, it became 'divergent' – the amount of deviation from normal flight would increase significantly with each Dutch roll cycle. The amount of deviation could double within 15 seconds.

Drag The force that resists movement of an aircraft through the air. There are three basic types: parasite drag, induced drag and wave drag. Parasite stems from the surface not being completely smooth or from obstacles in the air stream (e.g. the landing gear). Induced is a secondary effect of the production of lift. Wave is when shock waves are developed close to the surface of the aircraft in transonic and supersonic flight.

Dry bay Above each Concorde engine compartment there was an area called the dry bay, the purpose of which was to provide an area free of fuel and combustible fluids in which many sensitive engine controls and electronics could be housed. Each dry bay was separated from the engine housings by a heat shield.

Dutch roll A series of out-of-phase turns when an aircraft rolls in one direction and yaws in the other. It is uncomfortable to experience and can be divergent on swept wing aircraft at high, subsonic Mach numbers. Its name comes from the motion of a classic Dutch skating technique.

EADS European Aeronautic Defence and Space Company, forerunner of the merged European aerospace giant today known as Airbus.

Elevator A movable part at the rear of an aircraft that is used by the pilot, or autopilot, to rotate ('pitch') the aircraft about its lateral axis – the nose going up or down.

Elevon One of six movable parts at the rear of a Concorde wing that combine the effects of an elevator and an aileron. They are used by the pilot, or autopilot, to roll and/or pitch the aircraft about its longitudinal and/or lateral axes.

FBW *See* Fly-by-wire.

FDR (Black Box) A Flight Data Recorder (black box) records and stores digital data from a very wide number of aircraft sensors on to a medium designed to survive an accident. It can be used to reconstruct the nature and circumstances of a flight to help investigators determine the causes of an aircraft accident. It is housed in the most 'crash survivable' area of the aircraft and is painted orange – not black.

Fin The fin of an aircraft is usually a large vertical surface located on top of the rear of the fuselage. It improves vertical stability and contains the rudders on its trailing edge.

Fire Control Handle Alternatively called an Engine Shut Down Handle (ESDH) on Concorde, these were four controls, one for each engine, centrally located in the immediate overhead panel between the pilots. They contained a red light which illuminated if an engine fire was detected. The light was accompanied by the continuous ringing of a separate 'fire bell'. Pulling the relevant handle automatically initiated several actions designed to stop the engine, plus close off fuel, hydraulic and air supplies. When the ESDH was pulled, the engine fire extinguisher 'push buttons' became accessible.

Flame-holder A set of circumstances which can stabilise a flame against the flow of air and unburned fuel, so that it continues to burn. A change in one, or more, of the circumstances will cause the flame to cease.

Flaps Extendable, high lift devices on the trailing edge of the wings of most subsonic jet airliners. Their purpose is to increase lift during low-speed operations such as take-off, initial climb, approach, and landing. Because of the aerodynamic nature of its delta wing Concorde did not have, nor need, flaps.

Flight Crew The members of the crew directly responsible for aircraft operation and overall safety. The standard number aboard Concorde was three – two pilots and a flight engineer.

Flight Engineer The member of the flight crew directly responsible for aircraft systems but also, with the pilots, a fully integrated member of the cockpit team. Uniquely on Concorde, pilots' licences initially carried an endorsement saying that they were not valid unless the person at the systems panel was a fully qualified, licenced, engineer.

Flight Envelope A description, sometimes graphical, of the maximum speed, altitude, load, weight, centre of gravity, etc. for which an aeroplane was designed and/or can safely be operated.

Fly-by-wire (FBW) Semi-automatic, computer-regulated aircraft flight control systems that replace the conventional mechanical 'cable and pulley' system with an electronic interface. Concorde was the first airliner to use this system and, consequently, retained the mechanical system as a back up to its dual FBW controls which were designated 'blue' and 'green'.

Fowler flaps More complex flaps than ordinary flaps, they have a slotted design through which air can flow, allowing the wings to produce more lift at low speed.

Garda An Garda Síochána – the Irish national police service.

Glideslope The glideslope is the desired vertical descent path that a pilot aims to follow on final approach to the runway. It can be assessed visually, with the support of visual aids or within the

Instrument Landing System, where it is a radio signal which provides final approach vertical navigation guidance to sensors on the aircraft.

G Force and g G-force is a measurement of the type of force, typically acceleration, that causes our perception of weight. 1g is the nominal force on Earth – what we weigh. 2g would be experienced as twice what we weigh and 3g as three times, etc. – 'pressing us down into our seats'. Negative g would make us feel lighter than we weigh – 'floating out of our seats'.

GMT/UTC Greenwich Mean Time and Universal Coordinated Time are effectively interchangeable terms.

Great Circle Track The shortest distance between two points on the Earth's surface. If it is drawn on a conventional flat map it appears as a curve.

Ground-effect Technically, it's the reduced aerodynamic drag that an aircraft's wings generate when they are close to a runway. In reality, it feels like a 'big cushion' that Concorde sank into at around 50 feet above the ground. Generated by that massive delta wing it could almost totally stop the 850 feet per minute descent rate all on its own. Pilots just needed to learn how to handle it properly.

Hamble The College of Air Training where pilots between the ages of 18 and 26 were taught. Their residential 18-month course could take them from zero experience to becoming qualified CPL holders who would join either BEA or BOAC.

Handling Pilot The pilot, Captain or Co-Pilot, who is actually flying a particular sector and, in association with the crew, driving all of the decisions.

IAS Indicated airspeed – the speed of an aircraft as shown by its pressure sensitive instruments. It is an 'apparent' speed through the air because errors build as air density changes. The 'correct' speed through the air is known as 'true airspeed' (TAS).

ICAO The International Civil Aviation Organization, a specialised agency of the United Nations which is charged with coordinating and regulating international air travel.

IFR Instrument Flying Rules – rules and regulations to govern flight under conditions where the pilot's outside view is not available or safe. IFR flight depends upon flying by reference to instruments on the flight deck, and navigation is accomplished by reference to radio equipment.

ILS Instrument Landing System – a radio system at the ends of the runway that beams localiser and glideslope data to sensors on the aircraft. These inputs are displayed on the pilots' panels as the desired horizontal and vertical paths to be flown on final approach to the runway.

INS (Inertial Navigation System) An on-board pilot aid consisting of an inertial measurement unit (IMU) and a computational unit. The IMU is typically made up of a 3-axis accelerometer, a 3-axis gyroscope and sometimes a 3-axis magnetometer. Measurements taken from these enable the computational unit to determine the attitude, position and velocity of an aircraft, providing that it has been given an initial starting position and attitude.

Localiser The localiser is a radio signal which provides final approach azimuth navigation guidance to sensors on the aircraft. It is used to define the correct horizontal path towards the landing runway.

Mach Number & Mach The speed of an aircraft relative to the speed of sound. It is named after Moravian physicist and philosopher Ernst Mach. The speed of sound, in any prevailing atmospheric conditions, is Mach 1. Concorde cruised at Mach 2, twice the speed of sound or approximately 1,350 mph (2,170 kph).

Marmite factor Named after the UK's famously divisive beef-extract spread, which, reputedly, you either love or hate.

Mountain wave Significant 'wind rotors', or oscillations, on the downwind side of high ground resulting from the disturbance in the horizontal airflow caused by that high ground. Mountain waves can be very strong and cause downdrafts that are greater than an aircraft's ability to climb out of them.

Non-Handling Pilot The pilot, Captain or Co-Pilot, who is not actually flying a particular sector and, in association with the crew, monitoring the actions of the Handling Pilot and executing that pilot's instructions regarding the aircraft.

Normal Check List The pre-determined, trained, and practised sequence of events that must be followed sequentially to operate an aircraft from pre-engine start, through flight, to post-engine shut down.

NTSB National Transportation Safety Board – the US aircraft accident investigation authority.

Olympus 593 engine The Bristol (now Rolls-Royce) Olympus 593 engine was chosen to power Concorde. It was a variant of the Olympus engine that ships, electrical generating stations and earlier aircraft, such as the TSR-2 and the Vulcan, used.

P1, P2, P3, E/O P1 – commander; P2 – second in command; P3 – third pilot or systems panel operator; E/O – engineer officer.

Performance limited runway One where the prevailing atmospheric conditions reduce the maximum take-off weight 'on the day' to below the overall physical maximum take-off weight of the aircraft.

Piper Cherokee A four-seat all-metal, unpressurised, single-engine, piston-powered aircraft with low-mounted wings and a tricycle landing gear. It was designed for flight training, air taxy and personal use, and was the one chosen for intermediate flight training at the College of Air Training, Hamble.

Pitch The axis that extends crosswise over an aircraft (wingtip to wingtip) is called the lateral axis, and rotation about this axis is called 'pitch' – the nose going up or down.

PPL A Private Pilot Licence allows you to fly privately and for pleasure as the pilot-in-command of a single-engine piston aircraft, but not to be remunerated for doing so.

PSI Pounds per Square Inch – a measure of pressure

QE2 The *Queen Elizabeth 2* was designed for the transatlantic service from Southampton, UK, to New York. Subsequently she was also used in the cruise market. She served as the flagship of the line from 1969 until 2004. Following retirement, she has been repurposed as a floating hotel in Dubai.

RAF Brize Norton Situated in Oxfordshire, it is currently the largest UK RAF base and is primarily used for transport and training flights. It was frequently used for the training of Concorde crews and was the arrival airfield for the post-modification proving flight on 17th July 2001.

RAF Cranwell Situated in Lincolnshire, it is home to the RAF College where new officers and aircrew are trained.

RAF Fairford Situated in Gloucestershire, it is currently a 'standby airfield' but was the home of the UK Concorde flight test programme from 1969.

RAF Vulcan *See* Vulcan V-bomber.

Ramp The aircraft parking and manoeuvring area immediately adjacent to an airport's passenger terminals.

Reheat A long extension, at the back of an engine, which combines unburned oxygen with jet fuel that is shot into the high-speed exhaust from the engine's turbine and ignited. The additional thrust, in the region of 25 per cent, was used by Concorde for take-off and through the high-drag transonic region from Mach 0.95 to Mach 1.7.

Reverse thrust A mechanism that moves clam-shell-like doors at the back of an aircraft's engines to divert the thrust from propelling the aircraft forward to provide an element of retardation instead. The engines do not actually go into reverse, it is just the thrust that is re-directed.

Roll The axis that extends lengthwise through an aircraft (nose to tail) is call the longitudinal axis, and rotation about this axis is called 'roll' – the wings going up or down.

Rotors Wind phenomena, usually downstream of mountains, etc., caused by resultant large changes in wind speed and direction. They can lead to extreme near-surface turbulence, which can be very hazardous to aircraft. *See also* Mountain wave.

Royal Aircraft Establishment Based at Farnborough in Hampshire, it was a pre-eminent British research aviation establishment, famed for novel innovation and work on supersonic aerodynamics. It is now part of the UK Defence Research Agency.

R/T Radio-Telephony is the formal name of the radio system that pilots and ground personnel use to communicate with each other.

Rudder A movable part, usually on the fin of an aircraft, which is used by the pilot, or autopilot, to rotate ('yaw') the aircraft about its vertical axis – the nose going from side to side.

Runway threshold The first part of the runway surface that a pilot can use to land the aircraft on. It is usually, but not always, the physical beginning of the runway surface. It may sometimes be inset some distance down it because of a nearby obstacle on the approach path or to reduce noise in the immediate airport boundary area. In such cases there are distinct, specific, white line markings across the runway. New York (JFK) runway 31L is an example.

Seniority gradient The established, and/or perceived, seniority-based command and decision-making power hierarchy in a crew, and how it is balanced. Concentration of power in one person is a 'steep gradient', while more consultative and inclusive involvement of others is a 'shallow gradient'.

Shannon Airport Located in County Clare, Ireland, it is an international airport with a deep history of pioneering in global aviation. It was a busy refuelling stop for many international carriers in the 1960s, making it a gateway between Europe and the Americas. With the longest runway in Ireland, it is a major training airfield that was frequently used by Concorde, particularly as part of the 'Return to Service' programme.

Shanwick The name given to the 700,000 square mile Air Traffic Control (ATC) area of international airspace above the northeast part of the Atlantic Ocean. The name Shanwick is a blend of the words Shannon and Prestwick.

Shock cone or Mach one When an aircraft becomes supersonic the molecules of air in front of it form into a cone-shaped bow-wave – the shock cone. Its strength is dependent on the temperature of the air, aircraft design and weight and the speed and altitude at which the aircraft is flying. Concorde's was around 2.5 psi above the surrounding atmospheric pressure.

Shock cone or inlet cone An alternative method to Concorde's air-intake system to control supersonic air entering a jet engine. The Boom Overture will deploy this method on its engines, as did the SR-71.

Slats Extendable, high-lift devices on the leading edge of the wings of most subsonic jet airliners. Their purpose is to increase lift during low-speed operations such as take-off, initial climb, approach and landing. Without them, at those speeds, the aircraft would stall, as did Trident *Papa India* when they were inadvertently retracted in flight. Because of the aerodynamic nature of its delta wing Concorde did not have, nor need, slats.

SNECMA The *Société nationale d'études et de construction de moteurs d'aviation* (National Company for the Research and Construction of Aviation Engines) – the partner company to Rolls-Royce in the development of Concorde's Olympus 593 engines.

Sonic boom The perceived double-bang that an observer 'hears' when an aircraft passes at supersonic speed. It is a double-bang as shock waves are formed at both the nose and the tail of the aircraft. The bangs are what the ear perceives when the shock waves of plus 2.5 psi reach the ear drum.

SOPs *See* Standard operating procedures.

Speedbird The international radio callsign prefix for all BOAC and British Airways aircraft, except Concorde.

Speedbird Concorde The international radio callsign prefix for all British Airways Concorde aircraft. It alerted ATC, and other aircraft, to Concorde's extra speed.

SST An abbreviation for any type of commercial Supersonic Transport aircraft.

Stall A conventional aircraft stalls because the nose comes up too far, the speed is too low and the airflow generating lift over the wing breaks away. Lift then disappears, the nose falls and the aircraft pitches forward. Initially, the aircraft's flight controls may not function normally. With Concorde, at low speed, lift is generated by vortices, the strength of which increase as the nose rises, so its wing does not stall. But the increase in lift is eventually overcome by the greater increase in drag, and the aircraft descends – but its flight controls still function normally.

Standard operating procedures (SOPs) Within a large airline, flight crew may find themselves operating with individuals who they have never met before. SOPs provide them with a step-by-step guide to carry out flight operations effectively and safely. A particular SOP must not only achieve the task at hand but also be understood by a crew of various backgrounds and experience within the airline. In essence, your colleagues know what they expect you to do next or in a particular set of unusual or emergency circumstances.

Stick-shaker, wobbler and pusher A stick-shaker is a mechanical device designed to rapidly and noisily vibrate the control column of an aircraft, warning the flight crew that an imminent stall has been detected. If no corrective action is taken, it may be followed by a further device 'wobbling' the control column and/or physically pushing it forward.

Super-stall *See* Deep-stall.

Surge of an engine The compressor blades of a jet engine are airfoils not unlike wings. They will stall if the airflow is not maintained at the proper angle and/or rate. If that happens then the engine may 'backfire' and create a loud, audible 'bang'. The compressor may rapidly increase its rotation rate. Flames may also come from the back of the engine as the fuel/air mixture is now excessively rich due to insufficient oxygen. A Concorde surge at Mach 2 was a rare but very noticeable event. Corrective action included power reduction, computer changes, hydraulic power source changes and, if these were unsuccessful, descent to subsonic heights and speeds where recovery should be accomplished.

The Beast An occasional in-house nickname for Concorde, as was the 'Rocket'.

Touch and gos A training manoeuvre which joins a landing and a take-off together. After take-off, the pilot will fly a racetrack pattern to reposition on to final approach for a landing, followed by another take-off without bringing the aircraft to a full stop.

Transonic The high-drag region between subsonic and supersonic flight. For Concorde, that was from Mach 0.95 to Mach 1.7.

TSR-2 A British nuclear-capable supersonic Tactical Strike and Reconnaissance aircraft developed by the British Aircraft Corporation for the Royal Air Force in the late 1950s and early 1960s. It made its first flight from Boscombe Down on 27th September 1964. The project was cancelled by the UK government on 6th April 1965.

Tupolev Tu-144 A Soviet supersonic airliner designed by Tupolev. It first flew on 31st December 1968 and was fully retired in 1999. It operated a limited passenger service from 1975 to 1978 whereafter it was used as a cargo carrier and, latterly, a research aircraft. Its visual similarity to Concorde led to it being nicknamed Concordski, but it was far less sophisticated or successful than its namesake.

Turbofan A gas-turbine engine that uses its own fan for thrust.

Turbojet A gas-turbine engine that uses its exhaust to create thrust.

Turboprop A term that can denote a gas-turbine engine that drives a propeller, or shorthand for an aircraft that uses such an engine.

Type Certificate Issued by the appropriate nation's regulatory authority to a manufacturer, it is the approval of the design of a type of aircraft, and all its component parts. It signifies that the design complies with all the applicable standards.

Upper air work Higher altitude pilot and crew training on an aircraft or simulator. It would typically include items such as stalls, stall recognition and recovery, Dutch roll, high-speed operations and a practice emergency descent.

V_1 The pre-computed maximum speed in the take-off at which the pilot can take actions to stop the aeroplane within the available runway distance. The same speed is also the minimum speed in the take-off, following a failure of the critical engine, at which the pilot can continue the take-off and achieve the required height of 35 feet at the end of the runway.

V_2 Assuming an engine failure at V_1, and a continued take-off, it is the pre-computed target speed required to maintain a three-degree climb-out gradient, after clearing a nominal '35-foot obstacle' at the end of the runway. Concorde was much more sensitive to being climbed at the correct V_2 speed that conventional aircraft.

V_{MD} Minimum drag speed – the speed at which the aerodynamic drag on the aircraft is least. It is a function of aircraft weight. At maximum take-off weight, Concorde's V_{MD} was greater than its V_{MO}. At maximum landing weight, it was about 100 knots faster than the final approach speed. Hence, below about 265 knots, Concorde was 'on the back of the drag curve' where more power is needed to maintain a given lower speed.

V_{MO} The maximum operating speed which, for Concorde, varied both with height and weight. At sea level, it was 300 knots and at 50,000 feet it was 530 knots. As, at high weight, minimum drag speed was higher than V_{MO}, crews flew Concorde at V_{MO} whenever possible. Above 50,000 feet, speed was restricted to below V_{MO} by either the maximum Mach number or the maximum nose temperature.

V_R Rotation speed is the pre-computed speed in the take-off at which the pilot should initiate the rotation of the aircraft to the pre-computed climb-out angle.

VC10 A mid-sized, narrow-body, long-range British jet airliner designed and built by Vickers-Armstrongs and first flown at Brooklands, Surrey, in 1962. Designed to operate on long-distance routes from the shorter runways of the era, it commanded excellent hot and high performance for operations from African airports. In 1979, a VC10 achieved the fastest crossing of the Atlantic by a subsonic jet airliner of 5 hours and 1 minute, a record held for 41 years.

Vertical speed indicator (VSI) A flight deck instrument that shows the aircraft's rate of climb or descent in thousands of feet per minute (fpm). On most aircraft it is a circular dial but, on Concorde, it was a vertical strip instrument in order to cope with the potentially very high rates obtainable – up to 10,000 fpm climb and 30,000 fpm descent.

Vortex/vortices/vortex lift At a high pitch angle (angle of attack – 'alpha') spiralling vortices are formed at the tips of a conventional aircraft's high swept-back wing. On Concorde's delta wing these vortices were much stronger, more stable, and wider spread. These gave powerful 'high alpha lift' as they generated a high negative pressure field on the top of the wing. The physics effect is similar to that present in a cyclone-type vacuum cleaner.

Vulcan V-bomber Built by Avro, the Vulcan was one of three British bombers which made up the RAFs 'V-Force'. Nuclear-capable, the Vulcan had a delta-shaped wing, similar to Concorde, and was powered by an earlier variant of the Olympus engine.

VVIP A customer who was considered by an airline to be a 'very, very important passenger'. Other categories included VIP (very important passenger) and CIP (commercially important passenger).

Wear-strip A piece of the thrust reverser assembly of an engine, the wear-strip is designed to protect the part of the lower 'clam-shell' door which would be first to make inadvertent contact with the runway during a heavy, or high attitude, landing.

Yaw The axis that extends up and down through an aircraft (bottom to top) is called the vertical axis, and rotation about this axis is called 'yaw' – the nose moving laterally from side to side.

Yaw-damper A device to automatically limit an aircraft's undesirable, and uncommanded, yaw movements.

ZFCG The Zero Fuel Centre of Gravity is the aircraft's centre of gravity with everything except fuel aboard. Hence it is the centre of gravity of the basic aircraft adjusted for the specific location and weight of the passengers, crew, catering, baggage, cargo, etc.

ZFW The Zero Fuel Weight of an aircraft is the weight with everything except fuel aboard. Hence, it is the basic weight of the aircraft plus the weight of the passengers, crew, catering, baggage, cargo, etc.

1 Pitot head
2 Radome
3 Nose drooped position (17.5° down)
4 Weather radar scanner
5 Radar equipment module
6 Radome withdrawal rails
7 Radar mounting
8 Visor operating hydraulic jack
9 Pitot head, port and starboard
10 Visor retracting link
11 Retracting visor
12 Drooping nose operating dual screw jacks
13 Visor rails
14 Incidence vane
15 Front pressure bulkhead
16 Droop nose guide rails
17 Forward fuselage strake
18 Droop nose hinge point
19 Rudder pedals
20 Captain's seat, First Officer to starboard
21 Instrument panel, analogue
22 Internal windscreen panels
23 Overhead systems switch panel
24 Flight Engineer's station
25 Swivelling seat
26 Direct vision opening side window panel
27 Observer's seat
28 Circuit breaker panels
29 Avionics equipment racks, port and starboard
30 Starboard service door/ emergency exit
31 Forward galley units, port and starboard

32 Main entry door
33 Air exhaust vents, equipment cooling
34 Life raft stowage
35 Forward toilet compartment
36 Wardrobes, port and starboard
37 VHF antenna
38 Four-abreast passenger seating
39 Cabin window panels
40 Nose undercarriage wheel bay
41 Floor support structure above nosewheel bay
42 Nosewheel leg strut
43 Twin nosewheels, forward retracting
44 Spray suppressor
45 Nosewheel steering jacks
46 Telescopic rear strut
47 Nosewheel leg pivot mounting
48 Hydraulic retraction jack (2)
49 Retractable landing/ taxiing light
50 Ventral baggage door
51 Underfloor baggage hold
52 Forward passenger cabin, 40-seats in British Airways 100-passenger layout
53 Overhead light hand baggage rack
54 Cabin air duct
55 Passenger service units
56 Toilet compartments, port and starboard
57 Mid-cabin doors, port and starboard
58 Cabin attendant's folding seat
59 Stowage lockers, port and starboard
60 Fuselage skin panelling
61 Rear 60-seat passenger cabin
62 Cabin floor panels with continuous seat rails
63 Fuselage fuel tank roof panels
64 Conditioned air delivery ducting to forward cabin and cockpit

65 Fuselage conventional frame and stringer structure
66 Wing spar attachment double main frames
67 Starboard main undercarriage stowed position
68 Undercarriage bay central keel structure
69 Port main undercarriage wheel bay
70 Pressure floor above wheel bay
71 Rear cabin conditioned air delivery ducting
72 Cabin wall insulation
73 Cabin floor carried on links above stressed tank roof
74 Foot level cabin ventilating air duct
75 Cabin wall trim panelling
76 Dual ADF antenna fairings
77 Starboard main undercarriage pivot mounting
78 Inboard wing skin with tank access panels
79 Leading edge ventral Spraymat de-icing
80 Starboard wing main integral fuel tanks
81 Outer wing panel joint
82 Fuel/hydraulic fluid/air heat exchanger
83 Fuel/air heat exchanger
84 Engine fire suppression bottles
85 Starboard wing fuel feed tank
86 Fuel-cooled engine bleed air heat exchangers
87 Conditioning system cold air units
88 Engine bay heat shield
89 Outer wing panel integral fuel tank

90 Tank skin with access panels
91 Fuel pump in ventral fairing
92 Elevon hydraulic actuators in ventral fairings, fly-by-wire control system, electrically signalled
93 Mechanical trim and control back-up linkage
94 Dual outboard elevons
95 Starboard engine primary exhaust nozzle shroud
96 Combined secondary nozzles and reverser buckets
97 Starboard inboard elevon
98 Rear service door/ emergency exit, port and starboard
99 Cabin rear bulkhead with stowage lockers
100 Rear galley unit
101 Rear avionics equipment bays, port and starboard
102 Oxygen bottles
103 HF notch antennae
104 Fin leading edge structure
105 Multi-spar and light horizontal rib fin structure
106 Lower rudder hydraulic actuator
107 Upper rudder hydraulic actuator in starboard fairing
108 VOR antenna
109 Upper rudder segment
110 Rudder aluminium honeycomb core structure

111 Lower rudder segment
112 Extended tailcone fairing
113 Tail navigation light
114 Fuel jettison
115 Flight data recorder
116 Nitrogen bottle
117 Retractable tail bumper
118 Fin rear spar and tail bumper support bulkhead
119 Fin spar attachment joints
120 Rear fuel transfer tank
121 Fin spar support structure
122 Rear pressure bulkhead
123 Starboard side baggage door
124 Rear baggage compartment
125 Wing trailing edge root fairing
126 Port inboard elevon
127 Machined elevon hinge rib
128 Inboard elevon actuator in ventral fairing
129 Port wing rear main and feed integral fuel tanks
130 Machined wing spars
131 Inter-spar lattice rib structure
132 Combined tank end wall/nacelle mounting rib
133 Main engine mountings

Cutaway Illustration of Concorde

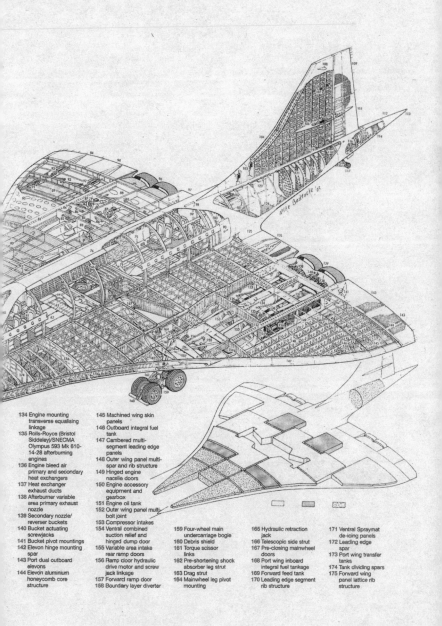

134 Engine mounting transverse equalising linkage
135 Rolls-Royce (Bristol Siddeley)/SNECMA Olympus 593 Mk 610-14-28 afterburning engines
136 Engine bleed air primary and secondary heat exchangers
137 Heat exchanger exhaust ducts
138 Afterburner variable area primary exhaust nozzle
139 Secondary nozzle/ reverser buckets
140 Bucket actuating screwjacks
141 Bucket pivot mountings
142 Elevon hinge mounting spar
143 Port dual outboard elevons
144 Elevon aluminium honeycomb core structure

145 Machined wing skin panels
146 Outboard integral fuel tank
147 Cambered multi-segment leading edge panels
148 Outer wing panel multi-spar and rib structure
149 Hinged engine nacelle doors
150 Engine accessory equipment and gearbox
151 Engine oil tank
152 Outer wing panel multi-bolt joint
153 Compressor intakes
154 Ventral combined suction relief and hinged dump door
155 Variable area intake rear ramp doors
156 Ramp door hydraulic drive motor and screw jack linkage
157 Forward ramp door
158 Boundary layer diverter

159 Four-wheel main undercarriage bogie
160 Debris shield
161 Torque scissor links
162 Pre-shortening shock absorber leg strut
163 Drag strut
164 Mainwheel leg pivot mounting

165 Hydraulic retraction jack
166 Telescopic side strut
167 Pre-closing mainwheel doors
168 Port wing inboard integral fuel tankage
169 Forward feed tank
170 Leading edge segment rib structure

171 Ventral Spraymat de-icing panels
172 Leading edge spar
173 Port wing transfer tanks
174 Tank dividing spars
175 Forward wing panel lattice rib structure

Index